Wuthering Heights on Film and Television

Wuthering Heights on Film and Television:
A Journey Across Time and Cultures

by Valérie V. Hazette

intellect Bristol, UK / Chicago, USA

First published in the UK in 2015 by
Intellect, The Mill, Parnall Road, Fishponds, Bristol, BS16 3JG, UK

First published in the USA in 2015 by
Intellect, The University of Chicago Press, 1427 E. 60th Street,
Chicago, IL 60637, USA

A catalogue record for this book is available from the
British Library.

Copy-editor: MPS Technologies
Cover designer: Stephanie Sarlos
Copy image: Outdoor Shooting: The Cameraman and A.V. Bramble,
A.V. Bramble and His Cameraman in the Stream.
    Courtesy of BFI Stills Department.
Production manager: Claire Organ
Typesetting: John Teehan

ISBN 978-1-78320-492-2
ePDF ISBN 978-1-78320-493-9
ePub ISBN 978-1-78320-494-6

Printed and bound by Short Run Press Ltd, UK

Contents

To Charles Barr, Mary Hammond and Ian W. Macdonald

List of Archival and Interview Material

Orientating Charts and Archival Material:

Interview Material:

Foreword

Liz Jones

Wuthering Heights has long fascinated audiences and continues to offer up its rich story for adaptation in a variety of media, including theatre, radio drama, music, the visual arts and, of course, film and television. Its admixture of a passionate and tragic love story, the struggle between the conscious and unconscious, power relations played out through class, sexuality, gender and race, together with heady doses of the Gothic and the Fantastic, makes for a potent and enduring brew. It is surprising, then, that *Wuthering Heights* has been relatively neglected within the field of adaptation studies; an oversight that this monograph goes some way towards redressing.

Spanning a period of almost a century, from Albert V. Bramble's 1920 silent film version to Andrea Arnold's 2011 adaptation, this fascinating study focuses on the wealth of film and television adaptations of *Wuthering Heights*. This hermeneutic approach creates a unifying focus on Hazette's intercultural, contextual and narratalogical explorations. While the ubiquitous UK (and joint UK–US) produced film and television adaptations are explored with freshness and clarity, Hazette also turns her attention to international film and television adaptations of the novel: from the classical Hollywood costume drama (William Wyler, 1939) to an oneiric, Balthus-inspired 1930s rural France (*Hurlevent*, Jacques Rivette, 1985); from a bleak, surrealist Mexican landscape (*Abismos de pasión*, Buñuel, 1953) to a luridly Gothic medieval Japan (*Arashi ga Oka/Onimaru*, Kiju Yoshida, 1988).

Wuthering Heights thrives on both the big and small screen. As a story it has also proved itself to be highly translatable to other cultures, as evidenced by those numerous transformations into other languages and cultures. As with adaptations, in general, each retelling does, of course, constitute an original and creative act in its own right. It is apt then that these case studies privilege the *creative* process of adaptation, while placing each adaptation in its wider historical and cultural contexts. Hazette also explores those deep, mythic structures that may be said to represent the ur-text of Emily Brontë's novel and that may, at least in part, account for its abiding (and intercultural) appeal. Illustrated with informative archival material, the book also includes interviews with film-makers in which they share some illuminating insights on the creative adaptative process. This engrossing study from a talented emerging scholar is sure to make a valuable contribution to the fields of both adaptation studies and Brontë studies.

Liz Jones is a Teaching Fellow at the Department of Theatre, Film and Television Studies, Aberystwyth University and Reviews Editor of the Journal of Adaptation in Film and Performance.

Introduction

Approaching a classic text through its film adaptations is similar to approaching it through its translations into a variety of languages rooted in many different cultures and times. The 'adaptation-translation' analogy, while laying the emphasis on multiple languages and destinations, allows literature, film and television to be placed on equal terms. This historical and transnational study of the film and television 'translations' of *Wuthering Heights* presents the afterlife of Emily Brontë's novel as a series of cultural journeys actualising the readers, film-makers and spectators-viewers as much as the films and television dramas themselves. In the course of this dynamic study, the cultural journeys, which are all interconnected, fleetingly reveal the transmission of meanings in and across cultures.

Wuthering Heights (1847) has inspired at least sixteen genuine film and television adaptations; this does not distinguish it, in principle, from such Victorian classics as *Oliver Twist* (1838) or *Jane Eyre* (1847).[1] What makes it an original text for this exploration into the immanence and diversity of culture relates to its extraordinary intertextual richness, which, in a nutshell, is expressed by its re-enactment of the Myth of *Psyche* and re-actualisation of the Romance of *Tristan and Iseult*, as well as marked by its profound allegiance to the *Lyrical Ballads*, first published by Wordsworth and Coleridge in 1798. This richness at the source shapes a complex textual identity that has spread its roots into the twentieth and twenty-first centuries. Acceding to the 'unconscious' of *Wuthering Heights* or, in the language coined by mythologist Claude Lévi-Strauss, to its 'constitutive units' or 'mythical components'[2] allows for a dynamic mapping of the text (Mythical Components and Bataillan Themes) condensed into a one-page Chart. Easily 'transportable' and 'printable', this Chart, which signals the importance of Part I dedicated to contextualising and methodology, will serve as a reference map throughout our *Journey Across Time and Cultures*, and feature, in its simplified form, before the Acknowledgements and Postscript. For signposting and mnemotechnical reasons too, significant terms may not merely be italicised but appear highlighted in bold 'when they are doing important work in the discussion'.[3]

Part II is dedicated to the lost picture of the British silent era (1920) directed by Albert V. Bramble and written by Eliot Stannard. The collaborative work of these two remarkable film professionals has seldom been researched. The recomposition of their *Wuthering Heights* is guided by the first two movements of '***trust***' and '***incursion***' that are described in the Hermeneutics of Translation of George Steiner, *After Babel* (1992), underpinned by a contextual study of the archival evidence (inclusive of related scripts by Stannard), and illustrated by the more striking pictorial elements of the film's advertisement campaign and shooting. This recomposition also firmly relies on the interpretation of some relevant biographical – Bramble's press biographies – and cinematic – Stannard-Hitchcock's *The Manxman* (1929) – material. The different clues thus brought to light stress the importance of Stannard-Bramble, the screenwriter-director team, in the practice of 'film adaptation' (or 'cultural translation'), and situate the lost picture's intertextuality in the incipient wave of Poetic Realism of the years to come. Further, the re-creation of the lost *Wuthering Heights* prompts us to relate the next cultural journeys, which the extant pictures will soon invite us to undertake, to an 'archaic' silent text that takes us on the Yorkshire moors as they were in Emily Brontë's time, and thus imposes itself as a source rather than a target text. As the divide between source-novel and target-film/target-television drama disappears, the unconscious of the text is ready to re-surface in the extant pictures, despite the weight of the dialogues and displacements characterising each of those transnational, heritage or 'cross-heritage' journeys.

In Part III, the discussion of the heritage and cross-heritage transformations opens chronologically with the study of two contrasting pictures, the Mexican version directed by Buñuel, *Abismos de pasión* (Paris-Mexico, 1933-1953), and the North American version directed by Wyler, *Wuthering Heights* (Hollywood, 1938-1939). These classic movies exemplify two different modes of film production and cultural translation. In the former, Buñuel, the *metteur en scène*, has immersed himself into the literary process of adaptation, and is closely associated with his cinematographer and successive screenwriters. *Abismos de pasión*, which deliberately twists the novel's mythical components, stages the Surrealist theme of *l'amour fou* by contrasting a disillusioned overtone with a profound imagery, and transforms the novel's 'initiatory love journey' into a 'fatal love attraction'. In the latter cinema classic, Wyler, the director, has missed the initial phase of cultural translation but finds strategies, with his chosen cinematographer, to *re-appropriate* himself a film that is greatly inflected by the screenplay dialogues, the decisions of the charismatic producer, Goldwyn, and the personae of the main actors. This well-remembered 'Olivier movie' – as it is often called – where Laurence Olivier plays Heathcliff, by tampering with the perception of Emily Brontë's *Wuthering Heights* and pretending to re-play the classic theme of *l'amour mercenaire*, gains the public's acceptance for a 'fatal love affair' thanks to the double-language of its cinematography. The cooler audience response to the later BBC adaptations (1967 and 1978), which reclaimed much more uncompromisingly the novel's dynamic structures and thematic complexity than the earlier ones (1953 and 1962), illustrates this argument.

Nourished by the positive energy of Anderson-Harris' unrealised *Heathcliff* (or *Love For Life*), which had already rippled through Leonard-Sasdy's Classic Serial (1967), the underrated British cinema movie honed by the Tilley-Fuest team (1970) inaugurated a punchy 'period drama' style rooted in the 'British New Wave'. This enduring 'period drama' style would be found much later in the Kosminsky-Devlin (1992) and Skynner-McKay (1998) versions, and was perfected to reflect the socio-historical truths depicted by Emily Brontë. The major innovation of the British period pieces of the 1990s was the enlargement of the cinema (and television) canvases to the second generation of lovers, which is at the heart of the novel's second volume. This generational enlargement was achieved without dismissing the compositional exactitude of the BBC2 *Classic Serial* by the Snodin-Hammond team (1978) but without resorting to its actual serial format. The Bowker-Giedroyc adaptation, created in 2009 for ITV, signified a departure from the social themes explored in the 1990s period dramas, and dropped the figure of Lockwood, the outsider. A contemporary take on the 1939 'fatal love affair', it stages a magnified Heathcliff, but uses *l'amour sensuel* as its dramatic engine. In the final discussion entitled 'The *Anti-Period Dramas* in Britain and Abroad', the limits to the adaptability of *Wuthering Heights* are being tested. If a British production could not stray from a Yorkshire setting, the three-episode serial of Wainwright-Sheppard's *Sparkhouse* (2002) showed that it could be turned deftly into an *anti-period drama* and refract the source text by swapping genders between Cathy and Heathcliff – and by de-multiplying the 'self-revelatory journeys'. The studies in cultural translation pursued by Rivette-Schiffman-Bonitzer in *Hurlevent* (1985), Yoshida-Bataille in *Onimaru* (1988) and Arnold-Hetreed in *Wuthering Heights* (2011) also broke the mould of conventional 'period drama' style. They took advantage of their cultural distance or, in the case of the latter production, of a culturally and ethnically distant Heathcliff, to remove the action to some foreign landscapes of the mind, and offer a compelling ending to this cycle of film adaptations, archival research and interviews. The cultural displacement effected by those last four productions may well have best '**compensated**' the text of the novel for the '**incursions**' of the film-makers, and allowed the optimal re-surfacing of the novel's dynamic structures and mythical components.

The most singular film and television translations of *Wuthering Heights* correspond to some privileged moments in interpretation and creation as the novel's more politically (and aesthetically) subversive components periodically ebb away and, in doing so, incessantly dwarf or magnify the dimension of the Other.

Part I

Contextualisation and Methodology

A

Contextualisation of Emily Brontë's Novel

Chapter 1

Myth, the Fantastic and *Wuthering Heights*

Since the invention of the cinematograph by the Lumière Brothers, the monopoly by fables, legends and novels of the power to stimulate individual imaginations into re-creating whole panoramas of events and characters has been lost. The externalisation of dreamlike images that had been carried by the whispering of archaic tales, or honed in the privacy of the readers' minds, became a technological reality in the side-shows of the fun fairs, then a marketable attraction on the big screens of the theatres where some avant-garde playwrights like Maeterlinck were reclaiming, for their symbolist plays, the mythical involvement of the spectators of Aeschylus' tragedies. A new social entity, the film audience, was born, and it would forever be fascinated by this ubiquitous mode of recounting of stories, all at once kinetic, mimetic and intangible. Michel Tournier, in *The Wind Spirit* (1977), exalts the formative role of myth[4] and images in the making of man. To him, the dazzling 'kaleidoscope of images' that 'surrounds and accompanies the little child from the cradle to the grave' needn't be a menace to her/his capacity to be 'creatively engaged with herself/himself and with the changing world around her/him':[5]

> L'homme ne s'arrache à l'animalité que grâce à la mythologie. L'homme n'est qu'un animal mythologique. L'homme ne devient homme, n'acquiert un sexe, un cœur et une imagination que grâce aux *bruissements d'histoires,* au *kaléidoscope d'images* qui *entourent le petit enfant dès le berceau et l'accompagnent jusqu'au tombeau.*[6]

As a young child, my imagination often came to life in the half-public, half-private glow of the television screen. In particular, I remember watching *La Belle et la Bête* (1946), and my profound excitement as I was drawn to its fantastic atmosphere. The mythical inspiration that fuelled the creations of Jean Cocteau, a translator of words into images, was breathing through this low-culture medium, the 'telly'. It was only in later life that the written word took on all its evocative force when I read a rare book that grasped my imagination, *Le Grand Meaulnes* (*The Wanderer*, 1913).[7]

Some years later, I was struck by the imaginative power of some dark tales, each of them interwoven with a number of 'Gothic' strands. I am thinking of Matthew G. Lewis' *The Monk* (1796), Mary Shelley's *Frankenstein* (1818), Edgar Allan Poe's *Tales of the Grotesque and Arabesque* (1840) – which connected with my earlier readings of the French texts, Villiers de l'Isle-Adam's *Contes cruels* (1883) and Guy de Maupassant's *Le Horla* (1887) – and of later works, especially Joseph Sheridan Le Fanu's *In a Glass Darkly* (1872) and Bram Stoker's *Dracula* (1897). Yet, above all these great nineteenth-century fictions belonging to the 'Fantastic' genre, Emily Brontë's *Wuthering Heights* (December 1847) remained conspicuously in my memory. Like Alain-Fournier's *Le Grand Meaulnes*, it can be easily written off as a humble *Bildungsroman*; its recommended reading, on the Continent, in the first years of an English Literature degree exemplifies its humbleness. Nevertheless, in the same way as the de-composition of *Bildungsroman* testifies to a rich underlying etymology with '*Bild*' meaning 'image' and '*Bildung*' meaning 'culture', the journey of discovery that constitutes *Wuthering Heights* is much more formative – and for a much wider audience, inclusive of cinéphiles, film reviewers and translators – than it may appear to be.

Embedded at the heart of *Le Grand Meaulnes* and *Wuthering Heights* stories is an extraordinary event that shapes the destinies of the young protagonists. There is the strange, bewildering children's party at the Domaine des Sablonnières where Augustin Meaulnes meets Yvonne de Galais, and starts a life of mysterious adventures and wanderings. There is the fateful night when young Cathy, allured by the lights of Thrushcross Grange, peers through the looking glass of the drawing-room window. At this very moment, she initiates a relationship with the Lintons, which is contrary to her attachment to Heathcliff, and wanders inexorably from Wuthering Heights, her true home.

The originality of *Wuthering Heights* lies in its complex narrative structure that includes the slightly differing recountings, by several narrators, of the same remarkable events. These events bend the destinies of the *sublime* heroes, Cathy and Heathcliff, who are deeply rooted in the Gothic layer of Emily Brontë's dark tale, towards tragedy and revenge. That is, until the *beautiful* heroes of the next generation, Catherine and Hareton, who embody a 'Wordsworthian' rather than 'Burkian' appeal to imagination, come onstage. Then, peace prevails, and the memory of the past tumult makes this peace fuller and much more significant for the reader. According to Claude Fierobe, the French specialist of the Irish Gothic novel,[8] Edmund Burke's treatise, *A Philosophical Enquiry into the Origin of Our Ideas of the Sublime and the Beautiful* (1757), represents a milestone in the genesis of the Gothic aesthetic:

> Whatever is fitted in any sort to excite the ideas of pain and danger; that is to say, whatever is in any sort terrible, or is conversant about terrible objects, or operates in a manner analogous to terror, is a source of the sublime; that is, it is productive of the strongest emotion which the mind is capable of feeling.[9]

Furthermore, Burke's pragmatic and unsentimental treatise inadvertently gave rise to the Romantic and pre-Raphaelite Aesthetics of Landscape, which is essential in feeling a continuity between the first (*sublime*) and second (*beautiful*) generations of lovers, between – in the words of literary critic David Cecil – the 'children of storm' and the 'children of calm'.[10] With its *sublime* and *beautiful* aesthetic components, its constant emphasis on boundaries and thresholds (especially the window motif) and a nested frame of narratives constructed as a network of gazes, *Wuthering Heights* undeniably belongs to the Gothic or, more generically, to the Fantastic genre as defined punctiliously by Todorov in 1970.[11] From the first chapter onwards, its realistic topography and historical backdrop are disrupted by uncanny encounters and strong emotional undercurrents:

> The Fantastic corresponds to a rupture in the recognisable order of things, to an irruption of the inadmissible into the immutable legality of our daily lives. It is not the total substitution of an exclusively miraculous world to the real world.[12]

*   *   *

Wuthering Heights also owes something to the oral tradition of storytelling and is therefore related to a form of prose fiction other – and older – than the novel, the 'Romance'. In the fourth essay of his *Anatomy of Criticism* entitled 'Theory of Genres', Northrop Frye explains that 'the popular demand in fiction is always a mixed form, a romantic novel just romantic enough for the reader to project his libido on the hero and his anima on the heroine, and just novel enough to keep these projections in a familiar world'.[13]

Emily Brontë's work could perhaps be termed not just as a 'romantic' novel but a 'balladic' one. Winnifrith and Chitham's well-researched biography on Charlotte and Emily Brontë (1989) describes how in the latter decades of the eighteenth century, their grandfather Hugh O'Prunty from County Down, Ulster, not only propagated radical political ideas but also retold sagas and sang ballads, by the light of his fire, in his little kiln cottage.[14] Reputedly, like a *seanchaí*, he would intensely involve himself in his stories, relying on facial expression and body language, pausing and hurrying where necessary, each time reworking, ever so slightly, tales that drew on archaic Celtic lore and pre-Christian myths. His dramatic devices aimed at creating a sense of suspense or immediacy, and he was probably resorting to kenning and hyperboles, as well as incorporating poetic and lyrical speeches within his prose narrative. As shown by Jean-Pierre Petit in *L'oeuvre d'Emily Brontë* (1977) and Sheila Smith in her essay, 'At Once Strong and Eerie' (1992), this mode of inventive recounting, and the very themes and dramatic devices characterising the ballad, is closely related to the narrative structure, content and style of *Wuthering Heights*.[15] The 'text' alone does not open all avenues of meaning, and the assimilation of *Wuthering Heights* to a 'Romance' – or, better, to a 'balladic novel' – establishes the essential connection between 'text', 'context' and 'subtext' – the 'uncounscious of the text'. The expression 'balladic novel' signifies concretely that a biographical, historical and sociocultural approach enriches the text of *Wuthering Heights* by bringing its mythical components and dynamic 'structures'[16] much more easily to the surface.

Amongst the international screenwriters and directors who used their imaginations for the film versions of *Wuthering Heights*, quite a few were originally taken by the novel's historical background, others by its idealised author and the particular 'Chinese box' narratorial structure – the Gothic Figure of '***Mise en Abîme***' – in which she may fugitively appear to engage. In due course, these very aspects informed their scripts and *mises en scène*. For Patrick Tilley who wrote – 50 years after Eliot Stannard – the script of the second British cinematic version (1969) and then for the director David Skynner (1998), the sociopolitical layer of the novel deserved to be given a great deal of visibility. For Peter Kosminsky (1992), the hovering presence of Emily Brontë materialised on to the screen in a cameo apparition and voice-over.

Chapter 2

Emily Brontë and Her Local Sphere

Emily Brontë was born on 30th July, 1818, in Haworth, and died there on 19th Dec, 1848. Haworth was then a large Yorkshire village of almost five thousand inhabitants, which constituted a major link between the moorland sheep-farms and the large towns of Bradford, Halifax and Leeds. As early as 1810, Haworth possessed several woollen mills of its own and produced almost as much wool as Bradford – and more than Leeds or Halifax. With the advent of mechanisation, the manual labour of thousands of families who had, for generations, been combing wool in their hill cottages became worthless. The first decades of the nineteenth century were turbulent times that saw Britain transforming rapidly into an industrial society. As it stood in the thick of agitation for the *People's Charter* and the *Factory Acts*,[17] Haworth was hit by some momentous social and political events. It was no longer buffeted only by the winds of the moors. Emily Brontë, during her youth, would have had to confront daily misery in the streets of her own village; starving weavers, including free labouring children, were compelled to seek work in the new textile factories. She also had easy access to the *Leeds Mercury*, a newspaper that, at the time, published several influential pieces by industrialist and political reformist Richard Oastler on child labour in the worsted mills of Bradford. A 'classic' amongst these articles was entitled 'Slavery in Yorkshire'.

Many writers before Emily Brontë had made a central figure of the starving orphan who climbs the social ladder, and thus *appropriated* the trend started, in the mid-sixteenth-century Spain, by the 'novelas picarescas'.[18] However, the powerful circumstantial evidence – gathered by the Winnifrith-Chitham team (1989) and by James Kavanagh (1985) – supports the conclusion that Winifred Gérin (1971) had reached in her earlier biographical investigation centred on Emily Brontë: Emily's inspiration for the orphan motif was deeply rooted in her local sphere of experience, and the foundling Heathcliff, stranded in Liverpool, was probably a very young victim of the Great Irish Famine.[19] Also, from the real-life persona of Emily Brontë's character, Heathcliff, there radiates what Frye calls a 'glow of subjective intensity' which is intrinsically more balladic (or 'Byronic') than novelistic, and quite atypical of a hero of picaresque literature.[20] By setting *Wuthering Heights* in a previous generation (1770s-1790s), Emily Brontë was, consciously or not, tracing the decline of the Yorkshire working farms and depicting, with total matter-of-factness, an archaic society and its uncivilised inhabitants. The overt violence of the latter, their fluidity of station, changeable identities and offensive bad language were attributes most shocking to a Victorian commentator like Elizabeth Rigby (Lady Eastlake). In total contradiction to the views of her editor J.G. Lockhart at the *Quarterly Review*, she wrote, in her article of December 1848, a lashing comment on the 'country squire'[21] and the well-born heroine of the fierce novel:

> [...] the aspect of the Jane and Rochester animals in their native state, as Catherine and Heathfield [sic], is too odiously and abominably pagan to be palatable even to the most vitiated class of English readers.[22]

James Kavanagh, in the introduction to his 1985 critical study of Emily Brontë, quotes Lady Eastlake's violent piece of criticism to illustrate how radical the novel was perceived by its detractors, and demonstrate how politicised *Wuthering Heights* actually is. Further, the realistic transcription of the Yorkshire vernacular that Emily Brontë invented, chiefly for old Joseph's speeches, testifies to her ability to achieve a kind of 'linguistic' mimesis with an other and lower stratum of society, while immersing herself into her local environment. For how could she have reached such a degree of empathy and preciseness in her re-creation of the local dialect by sticking to the familial sphere, and restricting her dealings to *one* local woman, Tabitha Akroyd, a domestic servant at the parsonage for many years? This is, however, what her sister, Charlotte, in her 'Editor's Preface to the New Edition of *Wuthering Heights*' (1850), seems to have strongly suggested when contending that 'though her [Emily's] feeling for the people round was benevolent, intercourse with them she never sought; nor, with very few exceptions, ever experienced'. The romantic legend of the reclusive genius thus began, and harmed Emily Brontë's reputation as an earnest and skilled writer. Nevertheless, Charlotte Brontë also carefully rounded off her previous sentence with those striking words, 'and yet she [Emily] *knew* them, *knew* their ways, *their language*' [emphasis added] which gives a particular resonance to the statement, redolent with admiration, of the Haworth dialect historian K.M. Petyt's:

> In the speech of old Joseph we have a remarkably detailed and accurate picture of the local dialect – and we must agree that Emily has coped surprisingly well with the recording of this, though her ear is untrained and she has no resources for transcription beyond the standard orthography.[23]

B

From the Novel's to the Films' Intertextuality

T he film and television adaptations we are going to travel with span almost the entire twentieth century (1920-2011). They sprang from the imaginations of a particular fraction of Emily Brontë's readership – the film- and television-makers. Their interpretative achievements illustrate how cinema and television, far from damaging the original novel, have celebrated across time a contentious specimen of the English literary canon, which is also an integral part of the world literary heritage. In our attempt to stop time briefly and come up with a synchronic tableau of the filmic and televisual transformations of *Wuthering Heights*, our *Journey Across Time and Cultures* needs to be anchored into the deepest layers of the intertextuality of the novel. To achieve this, the phrase 'cross-heritage', which originates in the analysis of a literary biopic, *Les Soeurs Brontë* (1979), itself featuring in a *French Film Directors* series dedicated to André Téchiné (2007), requires full contextualisation. Prior to examining closely the filmic text of *Les Soeurs Brontë*, Bill Marshall provides his readers with this rationale:

> *Les Soeurs Brontë* [...] belongs to an unusual corpus in French – and by extension European – sound cinema (the practice is standard, of course, in Hollywood, as we have seen), namely that of a historical or literary film set, and whose origin lies, in another country, in this case Britain. The most famous example is possibly Marcel Carné's *Drôle de drame* (1937), set in nineteenth-century London. [...] Given that much discussion of the heritage film of the 1980s has centred on national identity (Higson 2003, Austin 1996), it is intriguing to consider such examples of '**cross-heritage' movies** [...].[24][Emphasis added]

What is truly 'intriguing' is the presence of the legendary Brontë family, as well as the centrality of Emily Brontë's *Wuthering Heights*, in any literature-, cinema- or television-orientated discussions suffused with the whole idea of 'heritage' adaptations and, much more challengingly, of 'cross-heritage' transformations. From the traditional standpoint of the English critic Queenie Leavis, *Wuthering Heights* was indebted to two balladic novels by Sir Walter Scott, *The Black Dwarf* (1816) and *The Bride of Lammermoor* (1819).[25] Closer to us, English biographer Claire Tomalin, in her perceptive introduction of the 1978 *BBC Classic* serial for *Radio Times*, declared that she believed *Wuthering Heights* to be closely affiliated to James Hogg's dark Gothic tale, *Confessions of a Justified Sinner* (1824),[26] while, still closer to us, American academic Anne Williams put forward its next-of-kin relationship with the 'female' Gothic and, in particular, with Mary Shelley's

Frankenstein (1818), in the *Cahiers Victoriens et Edouardiens* (October 1991).²⁷ The subterranean layers of *Wuthering Heights*, however, must expand far beyond cultural and linguistic boundaries, if one recalls my own spontaneous superimpositions, in the opening chapter, of Jean Cocteau's film, *La Belle et la Bête* (1946), and Alain-Fournier's novel, *Le Grand Meaulnes* (*The Wanderer*, 1913), on to Emily Brontë's novel.

Wuthering Heights' subterranean layers are most clearly seen to function as a raw semiotic entity through the critical vista opened by English writer and film critic Martin Spence. In his 'Attempts to Film the Life and Works of the Brontë Sisters' (1987), Spence considers the 'striking resemblance' between 'the first part of [Emily Brontë's] *Wuthering Heights*' and the 'plot and theme' of two closely related movies by François Truffaut – *Jules et Jim* (1962) and *Les deux anglaises et le continent* (1971):

> And all along, Truffaut uses Brontë parallels to give direction to Roché's diffuse novel.²⁸ […] Throughout the film [*Les deux anglaises et le continent*], the emphasis is on *language and madness* (Emily Brontë) rather than the *sensuality and openness* to experience of Roché's novel.
>
> Although Truffaut always sought to deny that his films were superior to Roché's novels, in general they *wipe out inessentials to make the themes inescapable,* telescope with advantage, *translate* intellectual elements successfully into dramatic forms and interpolate very little. They are faithful renditions which shirk almost nothing.²⁹ [Emphasis added]

What are the novel's 'inescapable themes' and how do they relate to its mythical components and dynamic structuring?

Chapter 3

The Myth of *Psyche* and the Fairy Tale of *Beauty and the Beast*

The parallel reading of the legend of *Psyche*, a story belonging to the Greek mythology from which the 'domesticated myth' (or 'Fairy Tale') of *Beauty and the Beast* derives, can assuredly render *Wuthering Heights*' subterranean layers more easily accessible.[30] As explained by Anne Williams in her Wordsworth-inspired essay, 'The Child is Mother of the Man' (1991), the 'ghostly girl-child', Cathy, conjures up 'the West's most powerful myth of mind and soul', a founding myth retelling the perilous quest of the heroine, Psyche, for her other half, Eros. Williams further argues that this myth was first re-actualised successfully in the novelistic form with Ann Radcliffe's precursory Gothic tales, amongst which her still popular *Mysteries of Udolpho* (1794):

This female Gothic plot concerns a young woman alone in the world, whose inner life we share, experiencing the action from her point of view, and most often through her words. [...] she finds herself in some dangerous space, most often the house of a powerful and distinguished family [where] she undergoes a number of trials [...] She is attracted to [...] a mysterious man [...] [the hero/other, who], in the course of the action, generally undergoes a transformation from threatening to lovable and loving. In the end, the power of the 'supernatural' is mitigated [...] and the hero and heroine find happiness in marriage.

[...] in *Wuthering Heights* all the plot functions indeed *are* present: they are simply multiplied by two. There are two heroines named Cathy. There are two plots. The two houses have different (and opposite) functions relative to their two experiences: Thrushcross Grange is hell for the first Cathy, but a paradise for her daughter, whereas Wuthering Heights, a blissful Heaven for the mother, is a place of trial for the daughter. [...] Happiness may be found in either a physical or disembodied state [although] Heathcliff and Cathy's paradise regained is a version of earthly reality.[31]

Anne Williams' reading of *Wuthering Heights* tallies with Michael Popkin's conviction that 'the action in *Wuthering Heights* must cover two generations'. Popkin bases his brilliant hypothesis upon an intuitive rapprochement between Buñuel's *Tristana* (1970)/*Abismos de pasión* (1953) and Cocteau's *La Belle et la Bête* (1946)/*L'éternel retour* (1943), and validates it very convincingly through Bruno Bettelheim's interpretation of the 'original' Fairy Tale, *Beauty and the Beast*:

In *Wuthering Heights* the duality of existence is symbolized by the Heights and the Grange, and the 'artificially isolated aspects of humanity' are incarnated at first by Heathcliff and by Edgar Linton. [...] Cathy II transcends the opposition between animal and mind by educating an animal [Hareton], and the result of the synthesis she achieves is not only a mature view of things, as Bettelheim would phrase it, but an end to the original fairy tale as well. Her love for Hareton frees the original beast from his spell – Heathcliff looks into the eyes of the young couple and abandons his own monstrous plan of revenge.[32]

The Gothic and mythical components unveiled by the parallel readings of the Myth of *Psyche* and the Fairy Tale of *Beauty and the Beast* concern the intra-diegetic (or intra-narratorial) layers of Emily Brontë's novel. These components will be designated, in the 'Chart of the Mythical Components, Bataillan Themes and Planar/Gothic Figures', as, on the one hand, **First Mythical Component** – i.e., the correspondence between the contrasting houses (Paradise/Hell), which instantiates the connection between the Profane and the Sacred – and, on the other hand, as **Third Mythical Component** – i.e., the Initiatory Love Journey, which marks the passage from Destructive to Reconstructive Passion, or from Thanatos to Eros. In his synthetic 'Preface to the Original Edition of *Myths, Dreams and Mysteries*' (1956), the historian of religions Mircea Eliade establishes that it is in narrative form that myths come into being and endure, and demonstrates that it is the revelation of certain universal aspects of human experience, the revelation of the self, 'ontophany', and the revelation of the Sacred, 'theophany', which motivates the birth and re-generation of myths:

> Myths reveal the structure of reality, and the multiple modalities of being in the world. That is why they are *the exemplary models for human behaviour*; they disclose the *true* stories, concern themselves with the *realities*. But ontophany always implies theophany or hierophany. It was the Gods or the semi-divine beings who created the world and instituted the innumerable modes of being in the world, from that which is uniquely human to the mode of being of the insect. In revealing the history of what came to pass *in illo tempore*, one is at the same time revealing an irruption of the sacred into the world.[33] [Emphasis added]

Eliade's understanding of myths is akin to the translator's approach described by George Steiner in his most famous monograph, *After Babel* (1992). Both Eliade's and Steiner's strategies command the existence of the same pre-requisites: the timelessness and re-generation of myths (and classics) as well as 'the exemplary models for human behaviour' (Eliade) or 'the constancy of general human traits' (Steiner). In the seminal sixth chapter ('Topologies of Culture') of *After Babel*, Steiner clearly explains that:

Whether in speech or the plastic arts' [or in the cinematic and televisual arts], those myths and classics are re-told and presented with some superficial differences but profound similarities.[34]

It could therefore be judicious to apply Steiner's translator's approach to our synchronic review of the film transformations (or *translations*) of *Wuthering Heights*.

Chapter 4

Tristan and Iseult

S till, we have only partly elucidated, through *Psyche*'s Myth and its associated Fairy Tale, the relation between the novel's mythical components and its dynamic structuring of the Imaginary. The subterranean layers of *Wuthering Heights* also re-surface with a vengeance when looking back on the twelfth-century Romance of *Tristan and Iseult*, which is qualified, rather unsentimentally, as the 'one great European myth of adultery'[35] by Swiss literary historian (and moralist!) Denis de Rougemont in his 'inescapable' monograph, *L'Amour et l'Occident* (1939). The title was perceptively translated as *Passion and Society* (1939) in the United Kingdom, and a little more literally as *Love in the Western World* (1940) in its North American edition. In every edition and re-edition (1956, 1972 and 1989) of it, though, Rougemont goes on justifying the status of myth achieved by the recounting of *Tristan and Iseult*, without mentioning too much Iseult:

> The Tristan legend has many features indicative of a myth. First of all these is the fact that the author – supposing the legend to have had one, and one only – is entirely unknown. The five 'original' versions that have come down to us are artistic rearrangements of an archetype it is impossible to trace.[36]

Tristan and Iseult rapidly came to be associated with the thirteenth-century Arthurian Romance of Lancelot and Guinevere but retained its autonomy, for the continental Europeans, at least, owing to the influential opera, *Tristan und Isolde*, composed by Richard Wagner in 1865, and the re-edited Romance, *Le roman de Tristan et Iseut*, which its author, French medievalist and textual critic Joseph Bédier, successfully published at the beginning of the twentieth century.

However, are the dynamic structures of this literary archetype really impossible to trace? The images and themes that the readings of *Tristan and Iseult* crystallise have been carefully described by Anne Williams in her earlier essay on *Wuthering Heights*, 'Natural Supernaturalism' (1985). Her highlighting of the final tableau of the lovers' graves, a final tableau 'conclud[ing] almost all literary incarnations of the Tristan myth, providing the final evaluation of a *detached observer*', reveals the **Second Mythical Component** of our Chart, i.e., the connection between **the Living and the Dead**, induced by Imagination and strengthened by Dream Visions.[37] This final tableau helps in drawing the attention of the Extra-Diegetic (or Extra-Narratorial) Observer on to the three-dimensional Figure of *Mise en Abîme*, as well as in establishing the presence of **Lockwood** who is, within the text, the Substitute for all the Extra-Diegetic Observers, and becomes 'our' **Intra-Diegetic Outsider**. In the poetic epilogue to *Wuthering Heights*, Nature becomes in essence the soft wind, heath and hare-bells while, in the final tableaux of the various Tristan stories, Nature transforms itself into the intertwined hazelnut and honeysuckle (or into the simple briar plants) that make the lovers' graves one. By a powerful symbolic association, Nature is made to harmonise with the state of Mind of the '*detached observer*', who was once involved in the action (or **Diegesis**), but now considers with appeasement the end of the tragic cycle – and reaches out in her/his Imagination for a cathartic tableau that s/he could share with the Extra-Diegetic Observers (or Listeners-Readers-Viewers).

The **First Mythical Component** can thus be re-defined as a fusion between '**Mind and Nature**' which implies a humanisation of Nature and a naturalisation of the Supernatural (or of the Sacred) through Imagination and Poetry. This plays a major role in *Wuthering Heights*' epilogue, which is both realistic and lyrical, a memorable tableau that could have been sketched by Wordsworth and Coleridge for the purpose of their *Lyrical Ballads* (1798) and chosen to define Walter Benjamin's 'Auratic' perception:[38]

> During the first year that Mr. Wordsworth and I were neighbours, our conversations turned frequently on the two cardinal points of poetry, the power of exciting the sympathy of the reader by a faithful adherence to the truth of nature, and the power of giving the interest of novelty by the modifying colours of imagination.[39]

The breath-taking but inhospitable landscape of the moors, the sheltered country house of the Lintons (Thrushcross Grange) and the wind-swept farmhouse of the Earnshaws (Wuthering Heights), the mellowness of the summer breezes and the suddenness of the autumn storms, coalesce in a spectrum of subtly contrasting colours that depict the characters' inherited traits and changing moods – and, sometimes, reflect their true inner selves. Cathy's prophetic dream where 'the angels were so angry that they flung [her] out [of Heaven], into the middle of the heath on the top of Wuthering Heights, where [she] woke sobbing for joy'[40] – as well as Heathcliff's wilfully induced visions of her, eighteen years after her death, 'in every cloud, in every tree', 'in every object', 'in the most ordinary faces of men, and women', in her nephew Hareton, and in '[his] own features'[41] – testify to the power of the **Unconscious** (and of a trained **Imagination**) to transform topological and morphological landscapes into emotional landmarks, and to align **the Sacred** with Natural order **and** 'omnipotent **Love**'[42] – a classic variation on the **Third Mythical Component**, from Thanatos to Eros. Cathy and Heathcliff's love looms through *Wuthering Heights*' vertiginous construction of *Mise en Abîme*, and is also enriched by its balladic and mythical layers.

Lockwood's terrifying dream of Cathy appearing as a wandering ghost-child tapping at his window then clinging, with her little ice-cold hand, to his arm to be admitted in her former bedroom at the Heights[43] – as well as Ellen's fateful vision of Hindley as her 'early playmate', 'his little hand scooping out the earth' near the bottom of the sand-pillar,[44] a guide-post, in the middle of the moor, to the contrasting halls and to the village – indicate plainly that the Intra-Diegetic Outsiders undergo an initiation that makes them feel the strong correspondence between Liminal Places and Childhood, between **the Sacred and Childhood** – a variation on the **Second Mythical Component** that distinguishes *Wuthering Heights* from the conventionally Gothic.

Chapter 5

Georges Bataille and the Literature of Evil

Howing both *Psyche*'s Myth and *Tristan and Iseult*'s Romance, we still have not unveiled completely the novel's 'inescapable themes'. Literary critic and philosopher Georges Bataille, in his essay 'Emily Brontë' that features in his insightful monograph *Literature and Evil* (1957), shows how *Wuthering Heights* is characterised by its subversive theme of '*sovereign and hostile love*'.[45] In particular, Bataille flawlessly describes the rapports between the Sacred, Childhood and Evil, which, he argues, are at the heart of *Wuthering Heights*. In our final version of the 'Chart of the Mythical Components, Bataillan Themes and Planar/Gothic Figures', this distinctive leitmotif 'Sacred-Childhood-Evil' is divided into a twin set of Bataillan Themes, the Lovers' related themes (Interdicts 1, 2 and 3) extracted from Bataille's 'Emily Brontë', and the Outsiders' related themes (Trespassing-Learning 1, 2 and 3) that are not only noticeable on the narrative surface of *Wuthering Heights*, but also remarkably discernible on the 'textual bodies' of its different cultural translations on the cinema and television screens.

Cathy and Heathcliff lived a privileged childhood in harmony with Nature, at the frontier with the Sacred and enjoyed, free as they were from the decrees of Society, an Ephemeral Sovereignty. Any foreknowledge of this *sublime* stage – and, even more so, its wilful prolongation through remembrance and recounting – constitutes an infringement of the Christian Law and of the Law of Reason. Bataille's own views on the subject help in clarifying this first point:

> The two children spent their time racing wildly on the heath. They abandoned themselves, untrammelled by any restraint or convention other than a taboo on games of sensuality. [...]
>
> But society contrasts the free play of innocence with reason, reason based on the calculations of interest. Society is governed by its will to survive. It could not survive if these childish instincts, which bound the children in a feeling of complicity, were allowed to triumph. Social constraint would have required the young savages to give up their innocent *sovereignty*; it would have required them to comply with those reasonable adult conventions which are advantageous to the community.[46] [Emphasis added]

This **Ephemeral Sovereignty** is matched, at the level of Lockwood – and, consequently, at the level of the Intra- and Extra-Diegetic Observer(s) – by the desire to test the limits of new and hostile territories, and thus re-discover 'the domain of the Moment' experienced as a child. To reach for the Sacred and the **Present Moment** (**Interdict 1**) and start Learning involves an act of Trespassing or of 'temporary transgression' (**Trespassing 1**):

> The lesson of *Wuthering Heights*, of Greek tragedy and, ultimately, of all religions, is that there is an instinctive tendency towards divine intoxication which the rational world of calculation cannot bear. This tendency is the opposite of Good. Good is based on common interest which entails consideration of the future. Divine intoxication, to which the instincts of childhood are so closely related, is entirely in the present. In the education of children preference for the present moment is the common definition of Evil. [...] But condemnation of the present moment for the sake of the future is an aberration. Just as it is necessary to forbid easy access to it, so it is necessary to regain *the domain of the moment* (the kingdom of childhood), and that requires *temporary transgression* of the interdict.[47] [Emphasis added]

The unbearable tension between Morality and **Hyper-Morality** is felt most acutely by Cathy who dies for failing to reconcile her pledge of love to Edgar, her young husband, with her superior and anterior troth to her childhood friend, Heathcliff, who lives in a 'dream of a sacred violence which no settlement with organised society can attenuate'. And it is this gruelling confrontation between **Good** and **Evil**, embodied by Heathcliff, that the outsider Lockwood transmutes into a force that works at detaching him from some sterile social conventions linked to his over-protective milieu.[48] His unrequited romantic feelings for Cathy's daughter, Catherine, as well as progressive **Empathy** for Heathcliff's Plight (**Trespassing-Learning 2**), mirror the unfulfilled love between Cathy and Heathcliff as grown-ups, and their **Revolt against Good** (**Interdict 2**).

The **Third Bataillan Theme**, which is developed in its original form by Kiju Yoshida in *Onimaru*, aligns **Sexuality** with 'Immortality, but at the same time [with] **individual Death**'.[49] This is Yoshida's undisguised screenplay argument for Kinu/Cathy's delirious consumption after the birth of her daughter and double, Kinu/Catherine. In this temporary victory of Thanatos over Eros (**Interdict 3**), the lovers are either mythified or ostracised since society needs to retaliate against the offenders, especially modern and postmodern Western societies where a 'disinterested attraction towards death'[50] is considered 'Evil'. In *The Manxman*, the sanction takes a legal turn as Kate, straight after her suicide attempt, is brought to justice in a short but harrowing public session, presided over by the Deemster, who happens to be her lover Philip! In *Wuthering Heights*, when experienced from a safe distance by the Outsiders, the Third Bataillan Theme is not seen as a sanction but symbolises the **Life-Changing** – and not Life-Threatening – **Moral Transformation** that they undergo for their active participation in the archetypal tale (**Trespassing-Learning 3**).

The **First** and **Third Mythical Components**, and their associated **Planar Figure** of **Opposites** and **Doubles** and **Gothic Figure** of **Lineage** and *Mise en Abîme*, appear to be holding the mythical frame together. What happens when these dynamic structures disintegrate? Does the whole edifice of the film and television adaptations collapse? The edifice does appear to collapse in the early *Wuthering Heights* BBC teleplays (1948, 1953 and 1962). Due to their intrinsic lack in seriality and outdoor scenes, and programmed absence of revelatory passion, they have failed to render the story recognisable.[51]

As to the **Second Mythical Component** and its affiliated **Bataillan Themes**, their degree of visibility plummets when the Intra-Diegetic Observers do not undergo a profound moral transformation. Mirroring Lockwood's (and/or Ellen's) failure to evolve and keep up with the demands of the archetypal tale, the Second Mythical Component becomes fainter and fainter, and Heathcliff/Cathy's (and/or Hareton/Catherine's) love journeys dwindle into tales of social ambition, the so-called 'story of the stable boy and the lady' – a materialistic, myth-repellent scenario that George Bluestone sums up rather caustically in his chapter dedicated to the 1939 film version:

> A lady bred in respectable surroundings finds herself torn between one of her 'own kind' and an attractive stranger. […] [F]or romantic and sexual reasons, the lady clearly prefers the outsider. If the lady is not willing to run away with the stranger (as in Capra's *It Happened One Night*), she sets out to make him respectable, to improve his table manners as well as his income (as in Hitchcock's *Rear Window*). In *Wuthering Heights* [the 1939 film version], the story takes a tragic turn due to a run of bad luck.[52]

Figure 1. Chart of the Mythical Components (MCs), Bataillan Themes (BTs) and Planar/Gothic Figures – Intra-Diegetic Levels

	Lockwood or His Substitute(s), the Intra-Diegetic Outsider(s)	Cathy-Heathcliff Catherine-Hareton
First Mythical Component (MC1)	Correspondence between some contrasting topographical, botanical and atmospheric Settings, and the Characters' antagonistic traits and changing moods	Connection between human beings and awe-inspiring natural forces or beauties; Connection between the Profane and the Sacred
Second Mythical Component (MC2)	Dream Visions of Cathy as a ghost-child, of Heathcliff as a ghoul; Fantastic Visions and Liminal Places associated with Childhood	Revelatory Dreams and Imagination-Charged Visions; Connection between the Living and the Dead; Correspondence between Childhood and the Sacred
Third Mythical Component (MC3)	Initiatory attachment and Self-Revelatory Romantic Journey	Initiatory Love Journey and passage from Destructive to Reconstructive Passion, from Thanatos to Eros; *Psyche*'s Myth; Correspondence between Love and the Sacred
First Bataillan Theme (BT1)	Testing the limits of new and hostile territories; Re-Discovering the Present Moment Experienced as a Child (Trespassing-Learning 1)	Wilful prolongation of the Ephemeral Sovereignty of Childhood; the Present Moment and the Sacred (Interdict 1)
Second Bataillan Theme (BT2)	Workable tension between personal development and social constraints; Empathy with Heathcliff's Plight or with Evil (Trespassing-Learning 2)	Unbearable tension between Hyper-Morality and Morality; Revolt of Evil against Good (Interdict 2)
Third Bataillan Theme (BT3)	Undergoing a Life-Changing Moral Transformation (Trespassing-Learning 3)	Sexuality and issues of Physical Death and Re-Generation; from Eros to Thanatos and Back (Interdict 3)
Planar Figure	Rejection and Identification	Opposites and Doubles
Gothic Figure	*Mise en Abîme*	Lineage

C

Adapting the Adaptation Discourse to Our Corpus

Chapter 6

Adaptation, Translation and the Unconscious of the Text

Three Relevant F-Words: Fidelity, Foreignisation and Figure

An impressive amount of academic publications on film adaptation has cropped up since George Bluestone's ground-breaking *Novels into Film* (1957). There, Bluestone posits that novels and films have antagonistic rapports to words and images, and to language in general: this postulate has the major 'domino-effect' drawback of placing literature and film on unequal terms, of blotting out television as a privileged medium for the adaptation of classics, of eclipsing the transdisciplinary work of translators, and finally of forgetting all about Archetypal Criticism ('la Mythocritique') – as developed by literary theorists Northrop Frye[53] (1957) and Gilbert Durand (1960), and advocated by film adaptation scholars Dudley Andrew[54] (1984) and Julie Sanders[55] (2006).

Growing exponentially in the last decade of the twentieth century and, as regards UK academic journals, symbolically coming of age in 2008 with the removal of the 'slash' from ***Literature/Film Quarterly*** (turned ***Journal of Adaptation in Film and Performance***), the idea that Adaptation (and Media) Studies could rely no more on an enforced separation between words, images and performance coincided with the publication of an entire collection of essays and panel presentations cheekily entitled ***In/Fidelity*** (2008). There, the editors make a plea for 'a plurality of critical approaches (rather than the infinity of perspectives promoted by relativistic post-structuralism or the reductive and evaluative approach represented by near-absolute *fidelity criticism*)'.[56] Here is one of the most emblematic passages, excerpted from the thought-provoking transcripts of the presentations, which come across as casual in tone but are definitely not so in content, since they signal a critical minefield. This excerpt deals with the 'Persistence of Fidelity', sounds like a yellow-card warning for unruly contributors to Adaptation Studies – 'Fidelity Discourse: Its Cause and Cure' – and springs from Adaptation Studies 'guru' Thomas Leitch:

> Adultery is good; it's productive. I mean it's not good for the family, but it's great for the novelist. It can open up all these productive, newfound possibilities for writing fiction that we never had before. So instead of saying fidelity is good, infidelity is bad, why don't we say fidelity is maybe not so good and infidelity is better.[57]

In the context of our study of the film and television adaptations of *Wuthering Heights* (1920-2011), this deliberate cheekiness is refreshing but should not really come to us as a huge surprise. Leitch, after all, could nearly be seen as providing a moralising counterpoint to De Rougemont's 'fidelity-orientated' but, in a way, as subtly edifying *Love in the Western World* (1940), which, I now feel compelled to assert, was simply taken as a starting point to explore 'The Tristan Myth' – turned 'Tristan **and Iseult**' recountings in the labelling of B, Chap. 4 – as well as hint at the **intra-diegetic outsiders**, the early propagators-translators-adaptators of this 'unhappy mutual love', the troubadours from Provence and elsewhere.[58] As far as Leitch's (good) influence on contemporary Adaptation Studies is concerned, though, what is striking is the way his critical discourse has so swiftly reverberated throughout the academic writings of the best European scholars in the field. Accordingly, Katja Krebs finishes her wonderful introduction to *Translation and Adaptation in Theatre and Film* (2014) with a sustained metaphor (and repetition) fusing adaptation and translation as 'acts of love' while underplaying not the all-important 'phenomenology of performance', but the activist role that the Adaptation Studies 'troubadour-scholar' could fulfil. Just as there is a sociological aftermath for those trailblazing adaptations hit by corporatist (or even chauvinistic) reviews – Albert V. Bramble's and Andrea Arnold's come immediately to mind – a sense of loss and missed opportunity can often arise from a scholastic disengagement from the bustle of the 'main/most overt[/hidden] source[s] of *intertextual connection*'[59] or, in other words, from an **archaeology** of the hypermediatic-activist operations that promote the place of 'the Other' (or of 'the Foreign'). As noticed by Belgian academic Thomas Van Parys, whose review of David Kranz and Nancy Mellerski's *In/Fidelity* was published in Leuven-based journal *Image & Narrative* (2013), the notion of **fidelity** to the original/main source texts has not so much evolved in its critical significance and political implications but in the broader, more sophisticated definitions that are bestowed upon it:

> While the adaptations in this section [Chapters 4 to 7] have indeed been more tenuous, again the degree of infidelity in the case studies has stayed more or less the same throughout the book; it is mostly the authors' readings (and more importantly perhaps, their *definitions* of fidelity) that are changing.

> [...] infidelity is probably as much part of the discourse of filmmakers, reviewers and filmgoers, even if it is not phrased in such terms. In this respect, *discourse analysis* (rather than *fidelity evaluation*) would make up an essential part of the 'pathology' of fidelity Leitch proposes in his short presentation [...].[60] [Emphasis added]

This obsession with 'the pathology of Fidelity' has triggered a declension of 'F-words' including 'the Flawed taxonomies of Faithfulness'.[61] For this very motive, one should be careful about 'othering' the fidelity discourse when it appears to be symptomatic of an 'operational metaphor' **amongst others** and represent an 'operational mode', which, I agree with Patrick Cattrysse, should not be reduced to an 'operational norm'.[62] Lori Chamberlain in 'Gender and the Metaphorics of Translation' (1992) and Shelley Cobb in 'Adaptation, Fidelity, and Gendered Discourses' (2011) have carried out discourse analysis and started to study systematically the phenomenon of 'gendered metaphorics' in their respective, if amorously interrelated, fields. Their inspired approaches belong to an intellectual wave originating in such high-profile translation theorists as Antoine Berman (*The Experience of the Foreign*, 1984 and 1992) and Lawrence Venuti – the latter initiated the collection *Rethinking Translation: Discourse, Subjectivity, Ideology* (1992) – an intellectual wave that is deftly surfed by Translation and Adaptation Studies expert Márta Minier. In her perceptive article 'Definitions, Dyads, Triads and Other Points of Connection' as part of Krebs' *Translation and Adaptation in Theatre and Film*, she cites under Note 7 James Naremore's well-known assessment of the fidelity issue (2000):

> [...] most writing on adaptation as translation, even when it assumes a tone of quasi-scientific objectivity, betrays certain unexamined ideological concerns because it deals of necessity with sexually charged materials and cannot avoid a gendered language associated with the notion of 'fidelity'. [63]
>
> (Naremore)

> From an ontological point of view, it is illuminating that both translation theory and adaptation theory [...] use the *other* concept as a trope when elaborating on the nature of transformative practices of the kind.[64]
>
> ([Emphasis added] Minier)

Minier's diagnosis nicely complements Naremore's and, if we care to put metaphorical usage at the forefront, replaces an impossible prophylaxis against 'F-words' by an essential prophylaxis for 'the Other'. Towards the end of her development on 'Dyads in Translation and Adaptation Discourse', Minier brilliantly argues that the 'domestication/foreignisation' dyad which, incidentally, has many affinities with the last two movements of George Steiner's Four-Beat Hermeneutics of Translation ('**appropriation/compensation**') may well be 'immensely helpful' in the apprehension of large corpuses exploring interrelated specimens:

> Lawrence Venuti (2007, 2008), revitalizing Schleiermacher's related concepts (namely taking the writer to the reader and taking the reader to the writer), puts forward the much contested but, I feel, still *immensely helpful*, notions of foreignisation and domestication (the latter strategy may be perceived as of a more adaptative nature, if not as a synonym for 'adaptation' as rewording, rearticulation). Modes and degrees of retaining/re-introducing foreignness and [or] infusing a text with features recognizable as the receiver's own are valid fields of research – and strategies of practice – especially if one considers these attitudes, as Venuti later emphatically clarifies, as two ends of a *broad spectrum*. The fact itself that a translation is written in a language different from that of its (main but not necessarily only) 'source' is a

gesture of domestication in itself, even if the text otherwise abounds with signs of *otherness*.[65] [Emphasis added]

Furthermore, the 'gesture of domestication' that interlingual translation always foregrounds links up to the notion of '**cross-heritage** adaptation' that we have begun to define in the introduction to B, Chap. 3. There, Marshall's reflexion on the notion of **inter/national** film heritage is nourished by his rapprochement between Carné's *Drôle de drame* (1937) and Téchiné's *Les Soeurs Brontë* (1979), both set in nineteenth-century Britain, both insisting on a historical and literary reciprocal exchange (on the cinema screens) between the English and French cultures, and both matching the extended definition of adaptations as 'reinterpretations of established texts in new generic contexts […] or with relocations of […] a source text's cultural and/or temporal setting, which may or may not involve a generic shift', offered by Sanders in *Adaptation and **Appropriation**.*[66] *Les Soeurs Brontë* stages Irish actor Patrick Magee, acting in English but dubbed into French, in the role of Reverend Patrick Brontë and French narratologist-semiotician-literary theorist Roland Barthes, then at the height of his game, as William Makepeace Thackeray. Barthes was neither used to playing famous English novelists for the cinema, nor to composing a pseudo-philosophical piece written in the first person, like an autobiographical novel. In this only novel of his, *Fragments d'un discours amoureux* (1977), which he wrote as he felt compelled to assert the existence of the discourse of lovers, Barthes immediately sets out to explain his modus operandi:

Tout est parti de ce principe: qu'il ne fallait pas réduire l'amoureux à un simple sujet *symptomal*, mais plutôt faire entendre ce qu'il y a dans sa voix d'inactuel, c'est à dire d'intraitable. De là le choix d'une méthode 'dramatique', qui renonce aux exemples et repose sur la seule action d'un langage premier (*pas de métalangage*). On a donc substitué à la *description* du discours amoureux sa *simulation*, et l'on a rendu à ce discours sa personne fondamentale, qui est le **je**, de façon à mettre en scène une *énonciation*, non une *analyse*.[67] [Emphasis added]

Everything follows from this principle: that the lover is not to be reduced to a simple symptomal subject, but rather that we hear in his voice what is 'unreal', i.e., intractable. Whence the choice of a 'dramatic' method which renounces examples and rests on the single action of a primary language (*no metalanguage*). The *description* of the lover's discourse has been replaced by its *simulation*, and to that discourse has been restored its fundamental person, the **I**, in order to stage an *utterance*, not an *analysis*. [trans. Richard Howard]

His hyper-popular experiment in simulation thus gets rid of all the unwanted critical metalanguage, and there persists only his chosen 'fragments', which are governed by 79 '**figures**' (or 'gestures of the lover at work') such as 'Je-t-aime', 'Atopos' and 'Gradiva' (no need for the fastidious 'In/Fidélité' which, if one remembers well, does not belong to the 'primary language' of lovers):

La figure est cernée (comme un signe) et mémorable (comme une image ou un conte).[68]

In the course of our present critical journey, this '**figurative** structuralism' at work in *A Lover's Discourse* will have to be thematically inspiring. As **Psyche** transforms into a **Gradiva** walking with a lighter, surer, more actual transmedial gait, Sigmund Freud's *Delusion and Dream in Jensen's Gradiva* (1907) summons 'Textanalyst' Jean Bellemin-Noël's own translation and reading of *Gradiva* (1902), *Gradiva au pied de la lettre* (1983), where Wilhem Jensen's fictional character **Norbert Hanold** can, at long last, cease to be Freud's patient and become Emily Brontë's **Lockwood**. Barthes' successful novelistic experiment with *Fragments d'un discours amoureux* is indebted to Durand's **figurative** structuralism,[69] emerging with the publication of the first version of *Structures Anthropologiques de l'Imaginaire* (1960). It seemed to spell out that if I could safely head towards the same epistemological route in order to obtain a dynamic mapping of the *Wuthering Heights* source text (then discover the congruent **structures** and **figures** in the *Wuthering Heights* target texts), I could afford at no time the same elegant, metalanguage-free genre. As a writer and scholar straddling (critical) cultures often made to be antagonistic when they are essentially reciprocal, my ultimate aim is to showcase the inventive faithfulness of those differing yet connected adaptations by assessing the **transmission** of the dynamic structures and figures, with a particular emphasis on the dramatic usage of the Gothic architecture of ***Mise en Abîme*** and the degree of involvement (or 'foreignisation') of the extra- and/or intra-diegetic **observers-viewers**.

* * *

The success of Flemish scholar Patrick Cattrysse with his scientifically minded *Descriptive Adaptation Studies (DAS): Epistemological and Methodological Issues* (2014) was reflected by his presence as keynote speaker of the *Association of Adaptation Studies 9ᵗʰ Annual Conference* (September 2014), aptly christened 'Adaptations and Multiplicities'. Even before his seminal thesis *L'adaptation filmique des textes littéraires: le film noir américain* (Leuven University, 1990), Cattrysse had apprehended adaptation as translation and been pleading for 'a multiple perspective approach'[70] that would allow, for instance, the mapping out of those processes and devices that allow filmic narratives to travel better across nations and cultures.[71] His DAS, however, does not really take up the issue of transmission any longer since the author sets the limits of his huge contribution of 2014 to theorising Adaptation Studies (with a polysystemic accent on Target Text, Teleological Process and End Product) as well as offering an arsenal of well-wrought tools and approaches. Essentially, Cattrysse is leaving the field clear for his colleagues to experiment on 'large groups of adaptations' and thus deal empirically with the issue of **transmission**:

[…] the reader should not expect methodical in-depth analyses of *large groups of adaptations*. The focus will be on conceptual tools that serve the descriptive study of adaptations. I must leave actual application of the method to the talented researchers who come after me.[72] [Emphasis added]

While my obvious talent appears to consist in my ability to step on to a virtual minefield and pretend I am not afraid of detonating an anti-personnel mine, my real strength resides in my conviction that Cattrysse is obfuscating, because precisely of his **faithful adherence** to the multilateral '**PolySystem**' (PS) approach, the valid transmission-orientated '**Archetypal**' (or 'Textanalytical' or, I know, it sounds funny in English, 'Mythocritical') perspective. It is as if **Anthropological** or **Figurative** structuralism had never existed, and this cannot be right. My small-scale, flexible, mythocritical 'Chart of the Mythical Components, Bataillan Themes and Planar/Gothic Figures' is as much needed for a close-reading of the *Wuthering Heights* adapted texts as a translator's/hermeneutic model, capable of operating coherently on a large and varied corpus of (apparently unrelated) transmitted texts, is required for this transmission-focused mode of very fine research work. In her *Postmodern Mythology of Michel Tournier* (2012), Melissa Barchi Panek has recently demonstrated how valuable the Lévi-Straussian and, by ricochet, Durandian perspectives actually are for the epistemological study of interrelated contemporary texts:

> In the case of Michel Tournier's writings, the postmodern world, with its proliferation of mini-narratives, proves to be the driving force behind the transformation of the myth. It is my intention to examine on a structural level the mythical components of the stories and novels of Tournier in the spirit of Lévi-Strauss. In his 'Structural Analysis of Myth', Lévi-Strauss emphasizes the importance of including in one's analysis all the versions of the myth to find common traits and oppositions, as he states: 'The true constituent units of a myth are not the isolated relations but *bundles of such relations*.'[73]

As Lévi-Strauss explained a long time ago in his *Anthropologie structurale* (1958), through numerous dazzling case studies, there is no difference, ontologically speaking, between mythical thinking and scientific thinking, an argument that Gilbert Durand develops and substantiates further in his *Structures Anthropologiques de l'Imaginaire* (1960):

> Peut-être découvrirons un jour que la même logique est à l'oeuvre dans la pensée mythique et dans la pensée scientifique, et que l'homme a toujours pensé aussi bien.[74]
> (Lévi-Strauss)

> Comme le dit Jung , 'les images qui servent de base à des théories scientifiques se tiennent dans les mêmes limites... (que celles qui inspirent contes et légendes)'. Nous

soulignerons donc, à notre tour, l'importance essentielle des archétypes qui constituent le point de jonction entre l'imaginaire et les processus rationnels.[75]

(Durand)

Flirting with French 'Mythocritique' and the Dynamic Structures of the Imaginary: Gilbert Durand

Together with its Centre de recherche sur l'imaginaire (CRI)[76] and yearly journal IRIS, 'la mythocritique-mythanalyse-mythodologie' does not correspond to a fictional, moribund or isolationist type of critical approach, faintly echoing Archetypal Criticism. While it suffuses Barthes' *Fragments d'un discours amoureux* (1977), it is also an invigorating current in contemporary Comparative Literature.[77] More importantly to us, owing to its transdisciplinary roots in Bachelard, Jung, Lupasco, Minkowska and Piaget, it has fertilised such geographically distant and superficially unrelated fields as Textanalysis (and its Doppelgänger, Genetic Criticism) defined in translator Jean Bellemin-Noël's *Vers l'inconscient du texte* (1979),[78] and Sociocultural Anthropology as represented by translator Douglas Robinson in *Displacement and the Somatics of Postcolonial Culture* (2013). Robinson's somaticity/performativity dyad corresponds to a performance-orientated re-play of Durand's anthropological ***trajectory*** (or *anthropological dialectic*) that Durand characterises over several pages, with great heuristic precision, in his Introduction to *Structures Anthropologiques de l'Imaginaire*, as '*l'incessant échange qui existe **au niveau de l'imaginaire** entre les pulsions subjectives et assimilatrices et les intimations objectives émanant du milieu cosmique et social*'.[79]

This core definition, entirely typed in italics by Durand himself, in which he situates the continual exchange of the anthropological trajectory **at the level of the Imaginary** is often lost in its abridged translation into English, which conveniently brings out the subsidiary clause that immediately follows,[80] and does not mention the Imaginary at all. In order to neutralise this translative Flaw as well as 'Identify[ing] Common Ground' (2012) between the translingual discourse of the 'Imaginaire-anchored' SAI and the 'anti-F-word' rhetoric of Laurence Raw (who specialises in the study of the adaptation of British classics in Turkey and in the East at large), the explanatory side of Durand's definition of '**trajectoire** anthropologique' needs to be brought out in its intact yet familiar Foreignness, which involves developmental psychologist Jean Piaget's empirical view on adaptation as a three-phase process ('accommodation', 'assimilation' and 'représentation'):

Finalement *l'imaginaire* n'est rien d'autre que *ce trajet* dans lequel la *représentation* de l'objet se laisse *assimiler* et modeler par les impératifs pulsionnels du sujet, et dans lequel réciproquement, **comme l'a magistralement montré Piaget**, les *représentations* subjectives s'expliquent par 'les *accommodations* antérieures du sujet' au milieu objectif.

([Emphasis added] Durand's SAI, p. 38)

For the same purpose, the Robinsonian (and Bellemin-Noëlien) components of Raw's re-motivated definition of 'The Task of the Translator-Adaptator' need to be acknowledged, then confronted with the heuristic description of the dynamic, multi-layered structuring at work in the Imaginary – which is reflected in Durand's 'Isotopic Chart of Images'.[81] This dynamic structuring of the Imaginary has subconsciously nourished my 'Chart of the Mythical Components, Bataillan Themes and Planar/Gothic Figures of *Wuthering Heights*', and removed the (film and television) adaptations under scrutiny from their untenable position of unrelated artefacts, while guaranteeing the importance of the imaginative trajectory (or journey) of the extra- and/or intra-diegetic observer-viewer – and not merely that of the unique translator-interpreter, and '*no one else*':

The notion of *fidelity* can be treated as another *façade* as translators and adapters confine themselves to keeping to the spirit and letter of 'the original text'. [...] They should look *beneath the source-text's surface* to discover what they think is *its basic meaning*. By such *methods* the translator can *create* '*an imaginative construction*' of the source text that the translator – and *no one else* – believes truly represents the whole.

([Emphasis added] Raw (2012)[82] via Bellemin-Noël's Towards the
Unconscious of the Text, 1979, then Robinson's The Translator's Turn, 1991)

Le schème [ou la **figure**] s'apparente à ce que Piaget [...] nomme le '*symbole fonctionnel*' et à ce que Bachelard appelle '*symbole moteur*'. Il fait *la junction* [...] *entre les gestes inconscients de la sensori-motricité*, entre les dominantes réflexes *et les représentations*. Ce sont les schèmes [figures] qui forment **le squelette dynamique**, le canevas fonctionnel **de l'imagination**. [...] Les gestes différenciés en schèmes [figures] vont *au contact de l'environnement naturel et social* déterminer **les grands archétypes** *tels que Jung les a définis*. **Les archétypes** constituent **les substantifications des schèmes**.

([Emphasis added] Durand's SAI, pp. 61-62)

Nous entendrons par **mythe** un *système dynamique de symboles, d'archétypes et de schèmes [figures]*, **système dynamique** qui, sous l'impulsion d'un schème [d'une figure], tend à *se composer en récit*. Le mythe est déjà une esquisse de rationalisation puisqu'il utilise *le fil du discours*, dans lequel les symboles se résolvent en mots et les archétypes en idées. **Le mythe explicite un schème**, **ou un groupe de schèmes**.

([Emphasis added] Durand's SAI, p. 64)

Enfin cet isomorphisme des schèmes, des archétypes et des symboles au sein des *systèmes mythiques* ou de *constellations statiques* nous amènera à constater *l'existence de certains protocoles normatifs des représentations imaginaires*, bien définis et relativement stables, groupés autour des schèmes originels et que nous appellerons *structures*.

([Emphasis added] Durand's SAI, p. 65)

Pour l'instant, contentons-nous de définir **une structure** comme une **forme transformable**, jouant le rôle de protocole motivateur pour tout *un groupement d'images*, et susceptible elle-même de groupement en une structure plus générale que nous nommerons **Régime** *[ou Polarité]*.

([Emphasis added] Durand's SAI, p. 66)

There is no point, I believe, in giving a redundant linguistic translation to this heuristic picture of the anthropological structures of the Imaginary, which are 'dynamic' **and** 'transformable', and through which 'myth' acquires a high degree of definition. It might be good, though, in the context of this postmodern, transmedial study of *Wuthering Heights*, to shift the emphasis from 'anthropological' to 'dynamic' and rename the '**Anthropological Structures** of the Imaginary', '**Dynamic Structures** of the Imaginary'. This is to avoid, somehow, an unfortunate confusion with strict Lévi-Straussian (or Jakobsonian) structuralism (as opposed to Figurative structuralism) that was tackled by Durand in his tongue-in-cheek article, 'Les chats, les rats et les structuralistes' (1979) but is still in the air.[83]

What also seems necessary, or even compulsory, at this stage is to complement this heuristic picture grounded in theory with a heuristic sketch grounded in text, and pursue very succinctly the interpretative and structural work started in Part I B. With this in mind, I propose to confront an abridged version of our tailor-made reference Chart with Durand's heuristic labelling, by means of a quick re-reading of three key passages involving either one or the other of the two main intra-diegetic outsiders of Emily Brontë's *Wuthering Heights*.

Let us start with Nellie, and this revealing portrait of her as a character on the move on her own Imaginary trajectory, and inciting the extra-diegetic observers-readers to dive in the Gothic architecture of *Mise en Abîme* too. As I read the text, I fill this new, improvised, imperfect Chart:

I came to a stone where the highway branches off on to the moor at your left hand; **a rough sand-pillar**, with the letters W.H. cut on its north side, on the east, G., and on the south-west, T.G. It serves as **a guide-post** to the Grange, and Heights, and village.

The sun shone yellow on its grey head, reminding me of summer; and I cannot say why, but all at once, **a gush of child's sensations flowed into my heart**. Hindley and I held it a favourite spot twenty years before.

I gazed long at the weather-worn block; and, **stooping down, perceived a hole near the bottom still full of snail-shells and pebbles** which we were fond of storing there with more perishable things – and, as fresh as reality, it appeared that I beheld my early playmate seated on the withered turf; his dark, square head bent forward, and his little hand scooping out the earth with a piece of slate.

'Poor Hindley!' I exclaimed, involuntarily.

I started – my bodily eye was cheated into a momentary belief that the child lifted its face and stared straight into mine! It vanished in a twinkling; but, immediately, I felt an irresistible yearning to be at the Heights. Superstition urged me to comply with this impulse – supposing he should be dead! I thought – or should die soon! – supposing it were a sign of death![84] [Emphasis added]

It is our MC2 and the strong correspondence between Liminal Places and Childhood, between the Sacred and Childhood, which is very much at work here, or in Durandian terms, Nellie's Mystical Journey. Her journey is characterised by a wilful gesture of descent ('stooping down') towards her childhood with Hindley, and a progressive fusion between her sensory apprehension of the guide-post and her sensory apprehension of Hindley. Borrowing Durand's modus operandi and associating the symbol '*Cradle*' with this extraordinary 'hole near the bottom still full of snail-shells and pebbles' facilitates the connection of Nellie's journey with that of Cathy and Heathcliff, who have earlier been pictured huddled together under another related symbol 'the dairy woman's *Cloak*',[85] [Emphasis added] and even prepares the connection of her journey with the mystical ordeal of Heathcliff, as he strikes one side of Cathy's *Coffin* loose in order to fuse with his beloved in death.[86]

Let us continue filling this Chart with an episode where a cuddly 'mystical figure' for a change, that of the 'animal gigogne', curled-up cat Grimalkin, features but where Lockwood, after a positively disastrous initiatic nightmare,[87] is now predominantly responsive to his Heroic Trajectory:

Two benches, shaped in sections of a circle, nearly enclosed the hearth, on one of these I stretched myself, and Grimalkin **mounted** the other. We were both of us nodding, ere any one invaded our retreat; and then it was Joseph, **shuffling down** a wooden **ladder** that **vanished in the roof**, through a trap, **the ascent to** his garret, I suppose.[88] [Emphasis added]

There remains only Lockwood's episode of dramatic labour, buried at the beginning of his initiatic dream, which renders him similar, with his pilgrim's staff, to Cathy and Catherine in their own initiatic quests as the latter rides Minny, the horse, and the former uses pen and ink in the blanks of a 'Testament'.

The snow lay yards deep in our road; and, as we floundered on, my companion wearied me with constant reproaches that **I had not brought a pilgrim's staff**: telling me **I could never get into the house without one**, and boastfully flourishing a heavy-headed **cudgel**, which I understood to be so denominated. [Emphasis added]

Now that I have justified the use of my small-scale, flexible reference Chart, I need to find a translator's model, capable of operating coherently and complementarily on a vast and international corpus of film and television adaptations, which draw on the unconscious of *Wuthering Heights* at some different degree, through many creative processes, and in a diversity of contexts.

Figure 2. Improvised Chart of the Heroic, Mystical and Dramatic Structures –Intra-Diegetic Level of the Outsider

	Lockwood as **Intra-Diegetic Outsider**	**Nellie** as **Intra-Diegetic Outsider**
MC1 or Heroic Structure	Verbal Schème: to Discern Archetypal Epithets: Light vs Dark; Up vs Down Archetypal Substantives: Light vs Darkness; Heights vs Depths Symbol: the Ladder	
MC2 or Mystical Structure		Verbal Schème: to Fuse Archetypal Epithets: Deep; Hidden; Warm Archetypal Substantives: the Child; the Guide-Post with 'a hole near the bottom still full of snail-shells and pebbles' Symbol: the Cradle
MC3 or Dramatic Structure	Verbal Schème: to Reunite and Progress Principle of Rational Explanation: Causality Archetypal Substantive: the Pilgrim's Staff Synthème: the Initiation	

Chapter 7

From Film Adaptation to Cultural Translation

Adaptation can be envisaged as an ideologically charged journey during which archetypal structures and figures are dynamically translated and surreptitiously transmitted. From the philological perspective of George Steiner, who dedicated the whole of *After Babel* to cultural translation, and fleshed out the entire monograph over more than two decades (1975-1998), the choice of the translated literary text is often mythocritically motivated. Conversely, the translations of the chosen literary text, inclusive of the more immediatic, less hypermediatic ones, have an rejuvenating, mostly meliorative, effect on it:

> To class a source-text as worth translating is to dignify it immediately and to involve it in a dynamic of magnification (subject, naturally, to later review and even, perhaps, dismissal). The motion of transfer and paraphrase enlarges the stature of the original. Historically, in terms of cultural context, of the public it can reach, the latter is left more prestigious.[89]

Every film and television adaptation (or cultural translation) of *Wuthering Heights* that voices the novel's inherent political discourse and ensures the visibility of its dynamic structures pays a small tribute to the considerable achievement of a young female author who was herself re-interpreting, with some unusual creative force, those old motifs (or 'archetypes') prevalent in the recycling of myths (or 'Sacred tales'). The successful adaptation of Emily Brontë's text to differing cultural contexts identifies the process of film adaptation with an evolutionary yet archaic form of 'cultural practice', a positive phenomenon that was put at the centre of the debates by Mireia Aragay in her illuminating introduction to *Books in Motion* (2005), 'Reflection to Refraction: Adaptation Studies Then and Now'. There, as in Simone Murray's thought-provoking article, 'Books as Media: The Adaptation Industry' (2007),[90] one feels a definite commitment to set the course of adaptation studies towards history and media culture. Still, Aragay's choice of showcasing Dudley Andrew's seminal essay, 'The **Well-Worn Muse**: Adaptation in Film History and Theory' (1980), also suggests that she, unlike Murray, acknowledges the role of the 'Muse' in the *cultural practice* of adaptation[91] and is still responsive, in principle, to the critical works of the anthropologists of the Imagination:

> Adaptation [...] is a *cultural practice*; specific adaptations need to be approached as *acts of discourse* partaking of a particular era's cultural and aesthetic needs and

pressures, and such an approach requires both 'historical labor and critical acumen'. [Emphasis added]

The re-wording and re-imagining inherent in these particular 'acts of discourse' raises the specific themes of shared authorship and re-interpretation, of translation or 'transfer' from the pages of the novel to the film and television screens since translation is not limited to the narrow cross-linguistic field. This is how George Steiner, in the Preface to his book's emblematic second edition (July 1991), addresses and contextualises what Roman Jakobson had earlier termed 'intralingual translation' in one of his own seminal publications entitled 'On Linguistic Aspects of Translation' (1959):

> *After Babel* postulates that translation is formally and pragmatically implicit in every *act of communication*, in the emission and reception of each and every mode of meaning, be it in the widest semiotic sense or in the more specific verbal exchanges. To understand is to decipher. To hear significance is to translate. Thus the essential structural and executive means and problems of the *act of translation* are fully present in *acts of speech, of writing, of pictorial encoding* inside any given language. Translation between different languages is a particular application of a configuration and model fundamental to human speech even where it is monoglot. This general postulate has been widely accepted.[92] [Emphasis added]

The Hermeneutics of Translation put into practice by Steiner in 'Topologies of Culture', the sixth chapter of *After Babel*, does not limit its scope to a transfer of meaning(s) between two exclusively verbal sign systems either.[93] Steiner demonstrates, with a case study in *music* adaptation, that 'transfers' or 'intersemiotic translations' or 'partial transformations' generate and define culture itself:

> There is between 'translation proper' and 'transmutation' a vast terrain of 'partial transformation'. [...] This zone of *partial transformation*, of *derivation*, of *alternate restatement* determines much of our sensibility and literacy. It is, quite simply, the matrix of culture.[94] [Emphasis added]

After Babel and *Wuthering Heights*

Steiner's Hermeneutics of Translation faces up to the difficulty of mapping out 'some reasoned descriptions of processes',[95] and thus struck me as the ideal large-scale methodology to explore *Wuthering Heights on Film and Television* (1920-2011). I could hope to account for the diversity of, and correspondences between, the **heritage** and **cross-heritage transformations** that constitute the complementary legs of this *Journey Across Time and Cultures*.

Nonetheless, to envisage this study as one would envisage a dynamic ensemble of interrelated 'tableaux' could not have been achieved without adopting a combined *mythologist's* (Eliade–Lévi-Strauss–Durand–Tournier) and *translator's* (Steiner–Berman–Venuti–Bellemin-Noël) *approach*. This implies a profound belief in the 'timelessness' of myths and classics as well as asserts, in Steiner's already quoted words, 'the constancy of general human traits and, consequently, of expressive forms whether in speech or the plastic arts'.[96] This archaeological yet cross-heritage approach was not selected by Patsy Stoneman in her diachronic and heritage-orientated *Brontë Transformations: The Cultural Dissemination of Jane Eyre and Wuthering Heights* (1996), but contemplated in the superb volume of conference essays on the vast topic of 'Women's Writing in English as Part of a European Fabric' (July 2001) which she inspired, *Identity and Cultural Translation: Writing Across the Borders of Englishness* (2005). This type of approach was also contemplated, if not articulated mythocritically, around the same time by Kamilla Elliot in her incisive film theory-orientated chapter, 'Literary Cinema in the Form/Content Debate' (2003), which draws exclusively on *Wuthering Heights* and ends as follows:

> Further study is needed to probe the philosophical and semiotic issues in the depth and detail they warrant. The chief purpose of this chapter has been to show that *the form/content heresies of adaptation rhetoric and practice* so marginalized in the novel and film debate are central to its dynamics rather than peripheral and that they present difficult, challenging, and rich points from which to explore formal, contentual, and contextual elements of representation further in theoretically invigorating ways.[97]

Crucially, the fluidity in the station and identities of the protagonists (and observers) that was instilled by Emily Brontë matches the flexibility in the purely chronological sequencing of the narrative, which can be disrupted without entailing a loss in the recognition of the story; this strongly suggests a complementary 'mythodological' approach to the classic 'narratological' approach. Furthermore, the viewers who are situated on the brink, like the observers who live in the depths, of the abysmal Gothic structure that hallmarks *Wuthering Heights*, oscillate between Rejection and Identification, but are eventually drawn into the adventures of the Opposite/Double couples of lovers and enticed into a Journey of Discovery. It follows that the cinematic and televisual transformations of *Wuthering Heights* are particularly well suited to the implementation of the dynamic methodology of the 'mythologist-translator' – a 'discoverer' of liminal places and temporalities. To privilege flexibility of mind and wide sampling, I adhered to the **Fourfold Hermeneutic Motion** ('*trust*', '*incursion*', '*appropriation*' and '*compensation*') described and exemplified by Steiner in the fifth chapter of *After Babel*,[98] and validated within some complementary research contexts (via Schleiermacher and Benjamin) by the highly visible translators-scholars Antoine Berman (1984) and Lawrence Venuti (1995 and 2007) who both decisively argue in favour of the latter term of the 'domestication/foreignisation' translative-adaptative dyad. This reliance on the four-beat model of the hermeneutic motion allowed me to steer *Wuthering Heights on Film and Television* on its historical (or **archaeological**) and transnational (or **cross-heritage**) course.

*   *   *

The lost picture of the British silent era (1920) that occupies Part II invites the initiatory motion of '*trust*' (or 'élancement'), which is indispensable for this missing translation of *Wuthering Heights* to emerge under the good auspices of melodrama and pictorialism, at a time when the intertextuality of the novel is still palpable. For the production company Ideal, this ***hermeneutic of trust*** is rooted in *Wuthering Heights'* themes, characters and contrasting settings, and constantly sustained by the novel's literary status and legendary author(ess). This elusive picture also invites the motion of '*incursion*' executed, in the first instance, by the Stannard-Bramble team but refracted, incidentally, by film historian Ian W. Macdonald whose published articles on the 'silent screenwriter' and the 're-discovered scripts of Eliot Stannard' (2007 and 2009) trigger the interpretation of the written and pictorial evidence, while giving a solid scholarly basis to the imbrication of the relevant biographical and cinematic material. This ***hermeneutic of incursion***, initiated with Bramble's *Wuthering Heights*, will be prolonged and captured into the work of the Stannard-Hitchcock team on *The Manxman* (1929), a silent film that shares with Emily Brontë's novel some well-recognisable dynamic structures.

The re-creation of the lost picture of the British silent era takes up the entirety of Part II, and enables us to handle Part III, by establishing a mood of '***retrospective empathy***'.[99] This last point requires some explanation too. As I needed to re-imagine a film for which there was no footage, I was inclined to investigate its mode of marketing, production and exhibition, as well as evoke its forgotten makers and their extant films and writings. As I needed to recall the atmosphere of an epoch, mostly unfamiliar, I decided to magnify the setting of the Yorkshire moors. This strategy of **trust** and **incursion**, relying on a mixed *contextual/archaeological* and eventually *textanalytical/mythocritical* study, was deployed throughout this *Journey Across Time and Cultures* without obscuring, I hope, for the extant productions of *Wuthering Heights*, those significant *formal* and *contentual* 'elements of representation' that keep the 'Form/Content Debate' alive. Some primary source materials, such as the only extant still from the lost Albert V. Bramble movie, the seven photo-illustrated pages of Ideal's film programme (1920) and two rare location shots, are actually woven into Part II, while Balthus' Persian-ink drawings created for *Hurlevent des Monts*, and originally published in *Minotaure* (1935), only inhabit the two sub-chapters about *Hurlevent* (Part III B, Chap. 25) by their virtual presence. My own interviews with *metteur en scène* Jacques Rivette and some other contemporary practitioners facilitate this participative journey. All these original artefacts – archaeological or tailor-made – should help us, irrespective of who we are and where we come from, in fathoming then reviving in our own individual minds the *Wuthering Heights* adaptations, films or plays, that may otherwise have remained so culturally distant.

Figure 3. Brilliant 'Ideals' Coming (27th May, 1920)

Figure 4. 'Ideal's' *Wuthering Heights* (29th July, 1920)

Part II

The British Silent Era – Looking Back at a Lost Picture

A

Wuthering Heights and the Written Evidence

Chapter 8

Dealing with the Absence of Footage

Wuthering Heights
Directed by A.V. Bramble
Adapted by Eliot Stannard (1920, Ideal)
Featuring Milton Rosmer
6 reels; 6,200 feet
General Release on 25ᵗʰ Oct, 1920, UK

From the time I first delved into the *Wuthering Heights* of 1920, the absence of film footage presented an interesting challenge to study. This absence not only turned out to be an asset for the flexibility of mind it required, but also the empirical strength it would give to the treatment of the extant artefacts belonging to Part III, where film and television adaptation is understood as an inherited cultural practice.

Working without the 'raw material' of the film or tape is not unprecedented in the field of early cinema (or early television) studies. This is a difficult task for which some coping strategies have been developed, notably by Jason Jacobs in his exploration of early, unrecorded, television programmes.[100] Using written secondary material (an original programme, some publicity biographies and the reviews of trade papers), exploiting film-related visuals (a rare frame enlargement, some extant stills – from the shoot and the movie itself – and a local newspaper's photographs), drawing on some apposite footage (originating in the same director or screenwriter) and, finally, considering the findings of some fellow-researchers have all rendered it possible: seeing Emily Brontë's novel with the eyes of a 1920 audience, and recalling a forgotten and more personal style of movie-making.[101] A collage of filmic tableaux – if not an entire moving picture – where some of the motifs present in the unconscious of the novel can clearly be spotted, progressively acquires some form and significance.

For the only valid reason that the 1920 *Wuthering Heights* is an elusive artefact, quasi-impossible to grasp without spending a great amount of time on it, Albert V. Bramble's silent film has often been downgraded, then left for dead in the memories of its makers and audiences. Jacobs, in the introduction to his study of the lost television dramas, describes such problematic artefacts as 'ghost texts', while detailing the work that can be performed on them in order to revive them:

These are texts that do not exist in their original audio-visual form but exist instead as *shadows*, dispersed and refracted amongst buried files, bad memories, a flotsam of fragments. [...] My intention is the reconstitution or reconstruction, using this written material, of '*ghost texts*'.[102] [Emphasis added]

My own engagement with the lost film of *Wuthering Heights* is, necessarily, of a comparable nature.

Chapter 9

Ideal's Programme - Adaptation Seen as Cultural Practice and Commercial Strategy

This chapter makes use of records from the British Library Newspapers Archive, at Colindale in North London, and of a Film Programme by Ideal cured by the Brontë Parsonage Library in Haworth. The frame enlargement and two stills illustrating the Bramble-Stannard production are held by the British Film Institute Stills, Posters and Designs Department, London.

The first issue of the *Brontë Society Gazette* (May 1990) contained an appeal for help in locating the 'ghost' silent picture, the *Wuthering Heights* of 1920:

Little is known about this pioneering film and recent attempts to locate a copy of Bramble's *Wuthering Heights* have been unsuccessful. Michael Harvey, Curator of Film at the *National Museum of Photography Film and Television* at Bradford has embarked on a search for the film and an appeal is made to any members of the Brontë Society for information that may help us find it. If you think you can help please contact Jane Sellars, Director, Brontë Parsonage Museum. [Emphasis added]

To date, the appeal has had no success but, thanks to the *Brontë Parsonage Museum*, I was able to locate a ten-page programme originating in Ideal (the production company and distributor)[103] and dating back to 1920: permission was given to reproduce this programme in its integrality, which allows us to discover and interpret the printed pages of its texts and useful pictures.

The programme in question is a revealing fragment of the paraphernalia of on-site exhibition aids that were used to prime the audiences of silent films. On its front cover is laid out, at the top, the small lozenge-shaped logo of Ideal and, at the bottom, the daguerreotyped portrait of Milton Rosmer (as Heathcliff) framed in a medallion that is crowned by the big-lettered title 'Wuthering Heights'.

The names of Emily Brontë – '[Her] tremendous Story of Hate' – of the director, A.V. Bramble, of the leading actor (Rosmer) and finally of the writer, Eliot Stannard, are distributed on the sides of the medallion. The programme boasts five good-quality stills – 'Wuthering Heights (the farm)', 'Hindley Earnshaw returns', 'Heathcliff determines to leave Wuthering Heights', 'Edgar Linton & Cathy', 'The death of Cathy' – and a photograph of Thrushcross Grange (in reality, Kildwick Hall) taken by *The Yorkshire Observer*.

The possible merits of the film had been overshadowed when I became familiar with *The History of the British Film (1919-1929)* published in 1971, just a chapter of a book in the very substantial work of the pioneer scholar Rachael Low. In that, she heavily criticised Ideal for privileging the adaptation of Victorian novels with 'long complicated plots' and castigated its period pieces for their 'hired fancy dress appearance, executed cheaply and without imagination'.[104]

There would not have been any unwanted shadows had I discovered the Ideal programme prior to examining what Low wrote on the film, the production company and the director. If the melodrama of the acting is apparent in the stills featuring Heathcliff and Hindley, the novelty springs from the freshness of the outdoor scene with Edgar and Cathy, and the realisation that the movie was shot using some superb locations. The quality of the photography of the exteriors, and the careful topographical and historical reconstruction – the moorland scenery, the buildings, the old coach and horses – are such that the third-rate status granted by Low to a production, which might be melodramatic but is probably well-crafted, must be questioned.

Her *History of the British Film* was written well before the revolution that the FIAF conference held in Brighton in 1978 initiated for the scholars, historians, archivists and collectors eager to take the chance of trusting and exploring fully early British cinema.[105] Had Rachael Low viewed *Wuthering Heights*, it is uncertain that her modernly standardised, teleological vision would have adjusted to it, and actually enjoyed it.

When Low says that Ideal 'privileged the adaptation of Victorian novels', this is essentially true but far too reductive: all the profitable and renowned film companies (*Hepworth Manufacturing Company*, for instance) coveted such material. And Ideal, for its scripts, was not merely relying on 'Victorian titles' (borrowed from the novels of Charles Dickens, so well suited to cinematic transformation[106]) but also on more contemporaneous fiction, the 'best-sellers' of the time, typified by the novels of the prolific writer, Hall Caine – then immensely popular:[107]

An ambitious and unusually interesting list of new productions is announced by Ideal Films, Limited, which is maintaining the distinctive 'literary flavour' associated with its work in the past. [...]

'In our choice of stories, we have sought to strike a note of modernity', said Mr. S. Gilbert, Ideal's publicity chief, to a BIOSCOPE representative. 'We are appealing to the masses, and we shall endeavour to deal intelligently on the screen with those social problems which touch their daily lives, thoughts and aspirations.'[108]

The pages of advertising inserted in *The Bioscope* (and *Kinematograph Weekly*, the other prominent trade paper of the silent period), the film programmes handed out to the spectators in the film theatres and even the articles written in the non-specialised press are sufficient proofs of the high esteem in which literary adaptations were held – even those adaptations springing from novels with 'long complicated plots'. As Andrew Higson insists at the start of his chapter on Cecil Hepworth's *Alice in Wonderland*:

> [...] the adaptation of literary source-texts was one of the most important strategies by which English cinema in the 1900s and 1910s negotiated the development of more elaborate and extensive narrative films and attempted to attract a more upmarket, respectable audience and reputation.[109]

Chapter 10

Ideal's Programme - Gazing at *Wuthering Heights*, the Film

The ten-page programme put together by Ideal for *Wuthering Heights* is directed at this 'more upmarket and respectable' audience. Its first page, entitled CAST, offers some clues about the structure of the production; in this silent melodrama, the two generations of lovers are represented.

The six-reel *Wuthering Heights* can be converted into 90 minutes (approximately) of moving images, a length that was above the standard duration of the then average five-reelers, and would have allowed for the three Heathcliffs, two Cathys and two Catherines to feature in the printed list of the cast. Nonetheless, each cinema show of *Wuthering Heights* could vary slightly (or significantly) in length. It becomes clear after perusing Kevin Brownlow's *Sight and Sound* article, 'Silent Films: What Was the Right Speed?' (1980), that the duration of *Wuthering Heights* must have not only varied from one cinema to another but also, within the same cinema house, from one projectionist to the next:

> A 1915 projectionist's handbook declared – in emphatic capitals – 'THERE IS NO SUCH THING AS A SET CAMERA SPEED!' The correct speed of projection, it added, is the speed at which each individual scene was taken – 'which may – and often does – vary wildly'. And it declared: 'One of the highest functions of projection is to watch the screen and regulate the speed of projection to synchronise with the speed of taking.'

The CAST also offers some clues about the performance. The different epochs in the lives of the main protagonists (childhood, adolescence and adulthood) are marked by the casting of different generations of performers, ranging from an unknown child actor – Twinkles 'Twinks' Hunter, 'a London child' aged four and a half (Cathy) – to a stage professional, Mrs Templeton (Nelly Dean), who is the only character that remains stubbornly unchanged even when she should appear younger on screen and be faking Hindley's age.

> The reason why the number of artists taking part is so large is chiefly that the characters have to appear at so many different ages that it is impossible always for one person to play a part throughout. For example, the part of Heathcliff, the central figure, has to be played at the ages of 5, 12, 19, and onwards to 40.[110]

In the fourth chapter of her trailblazing *Brontë Transformations* (1996), Patsy Stoneman, who interviewed 'Twinks' in January 1995, tells us how deeply impressed Twinks had

"WUTHERING HEIGHTS"

CAST

❀

Heathcliff (age 5) (the gypsy foundling)	DERRICK RONALD
Heathcliff (age 12)	ALBERT BRANTFORD
Heathcliff (age 19-40	MILTON ROSMER
Mr. Earnshaw (who found and befriended Heathcliff)	CECIL MORTON YORK
Hindley Earnshaw (age 14) (his son and Heathcliff's enemy)	ROY LENNOL
Hindley Earnshaw (age 20-39)	WARWICK WARD
Frances (Hindley's wife)	AILEEN BAGOT
Cathy (age 6) (Hindley's	TWINKLES HUNTER
Cathy (age 13-24) sister)	ANNIE TREVOR
Hareton Earnshaw (age 15) (Hindley's son)	LEWIS BARBER
Hareton Earnshaw (age 22)	CYRIL RAYMOND
Edgar Linton (age 15) (afterwards Cathy's husband)	LEWIS BARRY-FURNISS
Edgar Linton (age 22-44)	JOHN LAWRENCE ANDERSON
Catherine (age 7) (his daughter and afterwards Hareton's wife)	AUDREY SMITH
Catherine (age 17)	COLETTE BRETTEL
Mrs. Linton (Edgar's mother)	DORA DE WINTON
Joseph (the Earnshaws' servant)	GEORGE TRAILL
Nelly Dean (age 18-54) (the Earnshaws' nurse)	Mrs. TEMPLETON
Attorney	EDWARD THIRLBY
Doctor	G. MALLALIEU
Revd. Mr. Shielders	ALFRED BENNETT

Figure 5. Ideal's Programme: Cast

"WUTHERING HEIGHTS"

FOREWORD.

"In 'Wuthering Heights' there is more adventure, more passion, more energy, more ardour, more love, than is needed to give life or fulfilment to twenty heroic existences, twenty destinies of gladness or sorrow"--Maeterlinck.

Emily Brontë wrote her great story at the early age of 29. A year later she died, leaving posterity to wonder what her mighty genius would have produced had fate spared her.

The late Mrs. Humphrey Ward traced the source of "Wuthering Heights" to the "Tales of Hoffmann." Others believe it to be found in a poem by Emily Brontë entitled "Remembrance." Dr. Wright, in his book, "The Brontës in Ireland," holds that it was in listening to her father's anecdotes of his Irish experiences that she obtained the weird material of "Wuthering Heights"; while Mr. Clement Shorter, in his well-known book on the Brontë family, says that it was partly her life in Yorkshire and partly the German fiction which she devoured that inspired the work. It is perhaps more artistically true to suppose that it had its origin in Emily Brontë's own untamed spirit, for as a piece of literature it stands all alone in its vehemence, its intensity, and in what Swinburne called the "dark unconscious instinct of nature worship" that streaked her passionate genius.

When "Wuthering Heights" was published it was received with a howl of indignation. As Emily's sister Charlotte Brontë has bitterly said, "too often do reviewers remind us of the mob of Astrologers, Chaldeans and Soothsayers gathered before the 'writing on the wall' and unable to read the characters or make known the interpretation." To-day the charactors stand clearly revealed and the book, in the words of Mr. Clement Shorter, "with all its morbid force and fire, will remain for all time as a monument of the most striking genius that nineteenth century womanhood has given us."

"Wuthering Heights" has been re-printed ten times in company with the works of Emily Brontë's sisters, and has also been re-printed seventeen times by itself.

The present picture was taken by Mr Bramble, the producer, in the locale in which the story is laid, and although Wuthering Heights and Thrushcross Grange, the two houses in and around which the action passes, are in ruins, the Old Hall, Haworth, the home of the Emmott family and a grand old building of the Tudor period, was kindly placed at the disposal of Ideal Films for the representation of the former, while for Thrushcross Grange, Kildwick Hall the beautiful Elizabethan residence of Mr. W. A. Briggs, was fortunately secured.

The most devoted labour was expended in order to secure absolute accuracy, and thanks are due in this connection to the great authority, Mr. Jonas Bradley, who lent his whole-hearted assistance in the production.

Figure 6. Ideal's Programme: Foreword

been by the Yorkshire moorland and how well she could recollect 'the dedication of the cast and crew' and their 'attention to detail'. She 'played the scene where Heathcliff first arrived – she remembers being asked to show resentment, which she did very thoroughly! – but the later childhood scenes seem to have been taken by Anne Trevor, who carried the part to adulthood'.[111]

The character of Lockwood does not appear in the cast, although he is one of the most prominent narrators in the novel, the other being Nelly Dean (aka 'Nellie' or 'Ellen'), and can become the perfect reader/viewer substitute considering his position of outsider in the story. Lockwood, in fact, does not show up before the 1939 movie by William Wyler, and the 1953 teleplay by Nigel Kneale. His disappearance from the screenplay is common enough, even in the more recent productions, while the writing-out of Isabella and her son, Linton, helps in obliterating the most unsettling parts of the tale, where Heathcliff cruelly mistreats his young wife, then terrorises his own son. Linton Heathcliff's blotting-out also has the advantage of forcing the plot into one direction only, the fast and inevitable pairing of the second-generation Catherine with Hareton.

The film, as recounted by Patsy Stoneman, carried an 'A' certificate 'which allowed children to watch it if accompanied by an adult'. Here is the opinion of Jonas Bradley, former headmaster at Haworth Primary School and advisor on *The Brontë Centenary at Haworth* (celebrated in 1917), on these departures from the novel, in an interview given to *The Yorkshire Observer* (5th May, 1920):

> Talking to so enthusiastic an admirer of the Brontë works and so full a repository of Brontë lore as *Mr. Jonas Bradley*, I found him warm in his admiration of the work of the company. He has had an opportunity of *studying the scenario*, and he praises the manner in which *the more gruesome elements of the story have been minimised*. 'They are hardly fit for children to know much about', he declared, 'and I think the scenario evades them very cleverly. *The ghostly episodes, for instance, are given a clever and quite justifiable interpretation.* I don't think the book will be in any way spoiled; on the contrary, I believe the film will make more people than ever read Brontë novels.' [Emphasis added]

This is more involvement in the scenario, more acknowledgment of the rational Gothic and more regional press coverage than you could expect, nowadays, during the pre-production and shooting of a literary classic!

*   *   *

After the CAST comes the FOREWORD, which proves to be a miniature literary review of Emily Brontë's *Wuthering Heights*. It starts off, as the preface of a novel, with a striking quote from the 1911 Nobel Prize laureate for Literature, the Belgian poet and symbolist playwright, Maurice Maeterlinck:

In *Wuthering Heights* there is more adventure, more passion, more energy, more ardour, more love, than is needed to give life or fulfilment to twenty heroic existences, twenty destinies of gladness or sorrow. [112]

Ideal's foreword deploys a wealth of literary clues and cultural details that will not be matched by the *Film Study Education Guide* designed for the star-ridden *Wuthering Heights* of 1992. After Maurice Maeterlinck, it is the turn of the 'late Mrs Humphrey Ward' and of 'Mr. Clement Shorter' to feature in the foreword of Ideal film programme. They were both prominent contributors to the Haworth edition of the Brontë's works – *The Life and Works of Charlotte Brontë and her Sisters*[113] – edited at the turn of the twentieth century. In her own introduction to the Penguin Classics edition of *Wuthering Heights* published exactly one century later, present-day academic Pauline Nestor regards Mary Augusta Ward's preface to *Wuthering Heights* – where the latter declares that 'Emily certainly owed something to Hoffmann's *Tales*' – as a harbinger of 'the changing tide of critical opinion', and gives it as much significance as the later pieces crafted by Charles P. Sanger (1926) or David Cecil (1934).[114]

Ideal, familiar with the connection that Mrs (Humphrey) Ward has established between *Wuthering Heights* and Hoffmann's *Tales* in her preface,[115] either unwittingly transforms – or spin-doctors a little – her scholarly discourse by printing, in their foreword, '*The Tales of Hoffmann*'.[116] Therefore, in the minds of the film programme's readers, Emily Brontë's novel probably comes to be associated with the ill-fated – but eventually tremendously popular – 'opéra fantastique' entitled *The Tales of Hoffmann* (1880) adapted by the French composer Jacques Offenbach from E.T.A. Hoffmann's supernatural *Tales*. This opera would remain, for a very long time, a tragic collective memory across all Europe, as a result of the horrific fire that devastated the Ringtheater, in Vienna, on the second night of its performance, there, in December 1881.

This reference to *The Tales of Hoffmann* signals the ambivalent affiliation of the British silent cinema, hovering between 'high' and 'low' culture. This ambivalent affiliation is heavily underlined by Christine Gledhill in her pamphlet, *Melodrama and Realism in 1920s British Cinema* (1991), and by Brian McFarlane in his essay, 'A Literary Cinema? British Films and British Novels' (1986):

For, contrary to popular film history, the cinema did not emerge in the twentieth century as a unique and original art. Rather it was formed in relation to existing art and entertainment forms, national cultural traditions and values.

(Christine Gledhill)

Just as in the 1880s Henry James promoted the novel as a serious art form by linking it with the then more prestigious art of painting, so film, once it was established as primarily a narrative form, turned to literature.[117]

(Brian McFarlane)

Chapter 11

Ideal's Programme - Gazing at *Wuthering Heights*, the Novel

S till, it is its privileged bond with the 'major' arts, the literary affiliation in particular, that Ideal wishes to publicise in their foreword, rather than its links with any contemporary theatrical 'sub-genre' (for example, the popular theatre based on melodrama), with the continental opera or with any entertainment form (such as the fairground or the music-hall). Therefore, Ideal foregrounds the canonical nature of *Wuthering Heights*, which most of the nineteenth-century literary critics had reluctantly acknowledged. The production company does so by citing an English poet who was both modern and of great renown, Algernon Swinburne – an avant-gardiste whose death, ten years before the film was shot, had helped extinguish the wrath of the moralising intelligentsia. He is the perfect supporter to cite next, with an excerpt from his perceptive 'Note' of 1877 that revolved around Charlotte Brontë, and preceded by a good few years his actual review on Emily herself in the *Athenaeum*:[118]

> There was a dark unconscious instinct as of primitive nature-worship in the passionate great genius of Emily Brontë, which found no corresponding quality in her sister's.

It is in this Coleridgean state of mind (and turn of phrase) that Swinburne chose to define Emily Brontë's creativity, pointing to one of the novel's most accessible Fantastic components: the wild, isolated and stormy situation of the Heights that lays bare the characters' personalities and, sometimes, briefly, reflects their inner selves.[119] Commercially speaking, it makes good sense for Ideal to turn to advantage the much debated elemental force of the novel ('its morbid force and fire' in the words of Clement Shorter) and, ahead of the screenings, praise the potential movie-goers who, undaunted, will come and see the 'accurate' film adaptation of such a formidable novel. Further, in tapping into the deep vein of the novel's critical history with even more gusto than into the vein of the legendary Brontë family, the distributor refrains from speaking the words of the common marketing speech, and equips the readers of the film programme with some kind of initiatory knowledge that would allow them to perceive more acutely the *sublime* aesthetics of landscape of *Wuthering Heights*.

Ideal's showcasing of *Wuthering Heights'* literary controversy culminates with a sentence taken from Charlotte Brontë's biographical notice to the 1850 edition:

Too often do reviewers remind us of the mob of Astrologers, Chaldeans and Soothsayers gathered before the 'writing on the wall' and unable to read the characters or make known the interpretation.

Chapter 12

Ideal's Programme - The Gender of the Author as Publicity Ploy
and Intertextual Evidence

Nnone of the production companies (or distributors) of the subsequent film adaptations can boast a similar degree of attention to the novel's early critical history in their advertising material. The emphasis on the gender of the author, which characterises Ideal's promotional campaign ('the most striking genius that nineteenth-century womanhood has given us', taken from the pen of Mr Shorter again) and, more recently, the Paramount British Pictures' campaign for the 1992 *Wuthering Heights* (where the charismatic singer, Sinéad O'Connor, was cast as Emily Brontë and narrated the entire story), is as audience-catching as it is well-founded. It defines a whole critical branch of English literature.

From what can now be construed as the dark ages of the Victorian era, dark ages when Emily Brontë and her sisters (or Mary Ann Evans) were left with no other choice than to publish their works under masculine pseudonyms,[120] *Wuthering Heights* comes out into the twentieth-first century as one of those multidimensional texts that still challenge literary conventions by inviting, by imposing even an open perspective. After Elizabeth Gaskell in 1857 and Mary Ward in 1899, Virginia Woolf applied her mind to *Wuthering Heights* (1916), and insisted on the validity – and on the potency – of the female thought:

> *Wuthering Heights* is a more difficult book to understand than *Jane Eyre*, because Emily was a greater poet than Charlotte. She [Emily] looked out upon a world cleft into gigantic disorder and felt within her the power to unite it in a book.[121]

Many post-Victorian intellectuals – amongst them some influential writers and literary critics, like Virginia Woolf herself – declared they abhorred cinema while being actually fascinated by the new possibilities that it was offering to artistic expression. In her 1926 essay, 'The Cinema', Woolf pictured it as an animal of prey devouring 'its unfortunate victim', literature,[122] curiously blind to its power of re-generation and enrichment, seemingly unaware of the striking similarities between adaptation and translation, since it is the 'same sequence of intuitive and technical motions which obtain in both [translation proper]'.[123] Can it be purely coincidental then that the first ever film version of Emily Brontë's novel was conceived in the late 1910s, under the bad auspices of many well-established writers, but under the aegis of modern fiction, at the dawn of gender criticism?

Wuthering Heights has, since then, elicited some enlightening discussions about myth, intertextuality and female aesthetics. Anne Williams' article, 'The Child is Mother of the

Man', in *Cahiers Victoriens et Edouardiens* (1991), and Dorothy Van Ghent's essay, 'On *Wuthering Heights*', in *The English Novel Form and Function Language Review* (1953), come immediately to my mind. Further, Emily Brontë's novel has sparked off the fascinating Japanese film version, *Onimaru* by Kiju Yoshida (1988), which does incorporate the second generation of lovers, and uses a round, moon-shaped mirror as the symbol of the reverse and complementary mother-daughter destinies – a feminist subtext often erased from the more predictable cultural translations of the novel into films.[124]

Chapter 13

Ideal's Programme - The Issue of Fidelity to the Novel through the Locations of Wuthering Heights and Thrushcross Grange

After having situated Emily Brontë's novel in a succinct, but punchy, literary review that highlighted its sensationalistic uniqueness and hard-won popularity, Ideal's foreword takes, in its final paragraphs, a more familiar path. It branches off towards the film ('The present picture was taken by Mr Bramble'), extols the merits of the original shooting location ('in the locale in which the story is laid') and lays the emphasis on the choice of historic mansions for Wuthering Heights and Thrushcross Grange:

> [...] the Old Hall, Haworth, the home of the Emmott family and a grand old building of the Tudor period, was kindly placed at the disposal of Ideal Films for the representation of the former, while for Thrushcross Grange, Kildwick Hall, the beautiful Elizabethan residence of Mr. W.A. Briggs, was fortunately secured.

The local newspapers truly revel in giving their Yorkshire readership details of the shoot. *The Yorkshire Post* (April 1920) and *The Yorkshire Observer* (May 1920) seize the opportunity to salute the local consultant on 'Brontë lore', Mr Jonas Bradley, who not only read the scenario but also helped the 'producer' – the 'director', in today's vocabulary – to find 'locations for the various scenes'. In the foreword's concluding sentence, Jonas Bradley is warmly thanked by Ideal for his active participation in the adaptative process itself:

> The most devoted labour was expended in order to secure absolute accuracy, and thanks are due in this connection to the great authority, Mr. Jonas Bradley, who lent his whole-hearted assistance in the production.

Keeping in mind that the logic of 'fidelity' to the novel's Yorkshire topography, which will prevail until Luis Buñuel's transposition of the action to a Mexican setting in 1953, was then such a fresh technological innovation flaunting the intrinsic possibilities of cinema itself, it is perfectly normal to run into the adjectives 'actual', 'authentic', 'faithful' and 'true' (or their equivalent) in every press article written, as early as April 1920, about the movie:

> [...] in order to prepare the correct atmosphere and background, the scenes are being enacted as nearly as possible at the actual spots mentioned in the novel.
>
> (*The Yorkshire Post*, April 1920)

imble and his assistant (Miss Murray) is to reproduce the story as
ι the actual scenes in which Emily Brontë set it. '*Wuthering Heights*
workshop', wrote Charlotte Brontë, and Mr. Bramble has gone to
' of the moors above Haworth in order that the film version of the
‚ ‚‚ ‚‚‚‚‚ De topographically authentic.

(The Yorkshire Observer, 5th May, 1920)

It is the charm of *Wuthering Heights* that one does not feel it to be a reconstruction of a bygone period. The Brontës would doubtless have marvelled, but we feel that they would have accepted it all as *a sincere attempt to preserve their genius for future generations by a different method of interpretation*. They would not recognise the Wuthering Heights and the Thrushcross Grange of the film, because the originals are in ruins, but they would realize how much the producer must have been helped in his work by those who keep the Brontë memory green. Thus, for Wuthering Heights the Old Hall at Haworth was placed at his disposal, while for Thrushcross Grange he obtained permission to use Kildwick Hall. And in addition, he had at his command the moors, the stone hedges, the streams which Emily Brontë knew and of which she wrote, and the result is *a series of natural settings which it would be difficult to improve on.* [...] *Wuthering Heights* is not merely a good film but it is *a faithful reproduction of the original*, and the two things do not always go together.

(The Times, The Film World, 2nd Aug, 1920)

The English backgrounds of Northland villages and bleak moors are absolutely true to the atmosphere of the book and beautiful in themselves [...].

(Kinematograph Weekly, Reviews of the Week, 5th Aug, 1920)

Together with the above piece from *The Times*' Film World, it is *The Bioscope*'s film review that seems to convey best the ageless nature, the mythical dimension of the setting. Although its critical vocabulary is not quite distinguishable yet from the vocabulary of a theatre review, this last piece (dated, like the previous film review from the *Kinematograph Weekly*, 5th Aug, 1920) betrays an appreciation of the film's adaptation process, of the cinematography and of the acting that is simply non-existent in the others. Here is the beginning of that eloquent review:

'*Wuthering Heights*' is a drama of character and atmosphere and of the interplay between the two. [...] The importance of the background, the atmosphere, in this wonderful story rendered it at once singularly well suited to the screen and a supreme test of the producer's ability. Whereas many novels stand to gain nothing and lose much on the film, '*Wuthering Heights*' presented a unique opportunity to realise in fact an effect which its author suggested – and could only suggest by words – in imagination. In seizing this opportunity, Mr. Bramble shows that unusual perception of scenic values which has marked all his work since 'A Nonconformist Parson' (1919). […] Removed from the setting which inspired it, 'Wuthering Heights' would have seemed an improbable story, and Heathcliff an inhuman character. […]

As Heathcliff, Mr. Rosmer gives what is certainly one of the most astonishing performances yet seen on the screen. In his portrait of this monomaniac racked by lust for vengeance, he rises to heights of emotional intensity seldom reached even on a stage, yet his effects are never exaggerated or unnatural. [125]

The common denominator between Ideal's foreword and the articles in the local newspapers (and trade magazines) is a general misconception relative to the actual location of the buildings from which Wuthering Heights and Thrushcross Grange are born. Neither of them, according to Winnifrith and Chitham's authoritative biography of Charlotte and Emily Brontë, originated in some old castellated hall (or fine country house) in the neighbourhood of Haworth.

This does not mean that the phonetics of 'Top Withens', an isolated Elizabethan farm with which Emily Brontë was well acquainted, built on the moorland to the west of Haworth, did not play its part in the creation of Wuthering Heights. Similarly, an arresting detail – the date 1801 – which starts the novel, was chiselled above the front door of Ponden House, a mansion also situated just beyond Haworth. It was one of the favourite destinations of young Emily and Anne Brontë, when they rambled on the moors, for its fascinating library and ancient family tragedies; no doubt the date on its front door tablet crystallised in Emily Brontë's mind. While Ponden could not possibly have served as the prototype for the impressive park at Thrushcross Grange, a derelict cottage serving as a shooting lodge during the grouse season – and some of its surrounding moorland – situated 'on the left-hand side on the road up from Ponden Hall', became the 'actual location of Wuthering Heights itself' in the BBC production of 1967. This old farm cottage and its adjacent land was then the property of a Mrs Bannister whose agreement producer David Conroy had been actively seeking to start shooting his serial. In the BBC internal message that was aimed at securing an insurance contract against fire on the Yorkshire grouse moor (and was signed by one of director Peter Sasdy's personal assistants, Pennant Roberts), this particular location is referred to as 'Ponden Kirk'.

It is in fact when she was teaching at Law Hill, North Halifax, during the academic year 1838-1839 that Emily Brontë discovered the setting for her contrasting pair of halls. This is confirmed, without much room for controversy, by the investigation of the Winnifrith-Chitham team and by a letter that I also uncovered at the BBC Written Archives, when looking for some clues about that same 1967 Classic Serial. Here is what Mr R.A. Innes, former Museums' Director in Halifax, felt compelled to write in a postscript, at the end of his answer to David Conroy, who wanted to obtain his permission for filming the outside of Shibden Hall – and part of its adjoining park:

I wonder if you are aware of a legend, or perhaps pious aspiration would be better, in this area. Emily, when she taught at a school some 2 miles from Shibden Hall, did in fact use Shibden Hall as Thrushcross Grange with its close topographical location. And what I think much more remarkable is her utilisation of a now demolished property called 'High Sunderland' for the exterior of Wuthering Heights.[126] (9th June, 1967)

In the late eighteenth century, Law Hill, the boarding school where Emily Brontë taught, was a big woollen farm on the high ridge to the east of Halifax. Back in 1825, it had been developed into the well-regarded school that she eventually came to know. Law Hill not only consisted of the former three-storey farmhouse that overlooked the so-called 'Shibden valley', but also comprised a number of outbuildings sheltering sheep and horses, cows and dogs, the proximity of which would be very much a factor in imagining *Wuthering Heights*.[127]

At Law Hill, Emily Brontë became familiar with the northern approaches to Southowram, through the Shibden valley, where she spotted a pair of notable local halls – High Sunderland and Shibden Hall. The castellated farm of High Sunderland, demolished in 1950, brooded remote and wind-swept below the brow of its northern hill, and presented the same external layout as the eponymous farm of the novel: buttresses and thick walls, as well as some very particular deep-set windows, divided by mullions.[128] Moreover, High Sunderland possessed an east chimney that would have collapsed into the kitchen fire, down below, in a violent thunderstorm, and a wealth of sculpture, both grotesque and pagan, reminiscent of the 'crumbling griffins and shameless little boys' described by Emily Brontë in Chapter I.[129]

By contrast, Shibden Hall, a mansion visible from High Sunderland's grounds that was sheltered by trees and surrounded by parkland, could make the perfect Imaginary template for Thrushcross Grange. Even the road junction at its rear, 'Stump Cross', would give Emily the phonetic blueprint for 'Thrushcross'. Further, in Emily Brontë's time, a disused church, Chapel-le-Breer, which perfectly merges with the image of the derelict, Gothic-like 'Chapel of Gimmerden Sough' – first appearing in Chapter III – existed at Southowram. It was situated near 'Sough Pastures'.

On the one hand, the twenty-first-century scholars, translators and film lovers, in order to free themselves from the oppressive feeling of a strictly Haworth-inspired *Wuthering Heights*, should be aware of the creative landmarks just described; these landmarks do not compromise the dualistic topography of *Wuthering Heights*. On the other, their critical minds cannot ignore that this type of biographical material stems from a mode of investigation dating back to the mid-1980s. With those nuances in mind, Ideal's approach still comes across as 'flawed', if only very subtly and wonderfully flawed. The qualities of the Ideal film programme's foreword remaining untarnished, we can relish in the experience of Steiner's first hermeneutic movement of **trust**.

Chapter 14

Ideal's Synopsis - Melodrama and Pictorialism

The seven-page synopsis featuring next in the Ideal film programme – and entitled '*Wuthering Heights*: Emily Brontë's Tremendous Story of Hate and its overthrow' – reads like a slow-paced, linear summary of key scenes, commented upon by a moralising voice-over. It is the magic of the visual aid (the five tableaux-stills[130] and the photograph of Thrushcross Grange, alias Kildwick Hall) combining with the written synopsis itself that must have filled the expectant spectators with a sense of deep curiosity for the most tragic episodes of Emily Brontë's novel, which are now performed, for the first time, on a cinema screen.

When comparing Ideal's synopsis, which is intended for the ordinary picture-goers, with the introductory paragraph of *The Bioscope*'s film review that gives the professionals of the cinematographic industry a film summary (August 1920), it is without surprise the former that reads less concisely and factually. The melodramatic and vigorous tone of Ideal is palpable in the voice of the third-person narrator while manifesting itself graphically, twice, with the capital 'Ps' that start the substantive '**P**ower'. Here are two supportive excerpts, one from mid-text, the other from the last paragraph:

> But, as the years wore on, a higher **P**ower intervened in this devil's game.

> And Heathcliff's face lost its hardness, and became beautified with hope and faith. [...] For the evil in him had perished utterly, and in all-conquering love – the love of the woman he had now rejoined – he had found the real **P**ower, and the only happiness.

Further, when superimposing the stills of the film (and photographs of the shoot) on to the close reading of Ideal's synopsis (or even on to *The Bioscope*'s film review), the imprint of melodrama appears clearly; but the joyous, breath-taking outdoors functions, in both novel and film, at another level of aestheticism. It is strikingly akin to the aesthetic of the *beautiful* that has often been epitomised by the 'host of golden daffodils' of William Wordsworth.[131] By contrast, the pivotal indoor scenes, especially those involving the first generation of lovers, belong to episodes of profound emotional intensity – such as Heathcliff's inevitable departure from the Heights, or Cathy and Heathcliff's last encounter before she dies – which match the *sublime* of Edmund Burke and, quite naturally, translate into melodramatic tableaux. The word 'tableau' is particularly appropriate here.

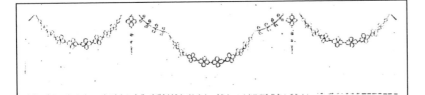

"WUTHERING HEIGHTS"

Emily Brontë's Tremendous Story of Hate and its

overthrow.

SYNOPSIS.

ONE day, genial Mr. Earnshaw, master of Wuthering Heights, came home from a walk on the lonely Yorkshire moors with a strange "find"—a ragged gipsy boy. He installed him in the house as though he had been a child of his own.

But by Earnshaw's son, Hindley Earnshaw—a cruel, mean, bad-tempered lad of fourteen—the newcomer was received with scornful dislike, which presently deepened into intense hatred. He bullied and beat the helpless little stranger until Mr. Earnshaw, in despair, banished the brute to college.

His pretty little tom-boy sister, Cathy, however, made of Heathcliff, as the little fellow was called, a playmate. Together they roamed the moors, and Heathcliff became the slave of Cathy's every whim.

Time passed, and old Mr. Earnshaw paid the debt of nature. Hindley, now married, returned, master of Wuthering Heights and all its domain, and determined to gratify his lust for vengeance on the innocent cause of his exile. He abused him, and degraded him to the status of a farm-hand, so that Heathcliff grew to be dirty, uncouth, sullen and brooding. But he smothered the rising mutiny within him, clinging with growing passion to his now fully awakened love for Cathy.

Figure 7.

"WUTHERING HEIGHTS"

"WUTHERING HEIGHTS."

One day, she tempted him to follow her on to the moors, where both were lost. The next morning the weary couple were found by young Edgar Linton, of Thrushcross Grange, who escorted Cathy to his home. Heathcliff, returning to Wuthering Heights, was mercilessly thrashed for his presumption in keeping company with his master's sister.

Seven years passed by. Cathy had blossomed into a beautiful but capricious woman of twenty. Hindley Earnshaw, now a slave to drink, had sunk into a state of degradation, cursing his baby boy, Hareton, in giving life to whom his wife had forfeited her own. Heathcliff, rough and loutish, loved Cathy with increasing ardour, but when one day he overheard her confessing her preference for him, but avowing her horror of marriage

Figure 8.

"WUTHERING HEIGHTS"

HINDLEY EARNSHAW RETURNS, MASTER OF
"WUTHERING HEIGHTS."

with one who had been brought so low, her words stung him sharply, and, mounting a horse, he disappeared moor-wards.

But on that day deep hatred was born in him.

Four more years passed, and Heathcliff returned, a changed man—handsome, well-groomed and wealthy. But during his absence Cathy had given her hand to Edgar Linton, and the discovery turned Heathcliff's hatred of Hindley into a savage obsession. He took advantage of his enemy's newly developed gambling habits to play him for his money and his lands—and he won. His love for Cathy, whose feelings for Heathcliff had never been quenched, led to stormy scenes with her and her husband, and when, under the stress of many terrible

Figure 9.

"WUTHERING HEIGHTS"

HEATHCLIFF DETERMINES TO LEAVE "WUTHERING HEIGHTS."

experiences, she died in childbirth, it was with Heathcliff kneeling, frantic and maddened, beside her deathbed.

Deeper and deeper sank Hindley Earnshaw in his debased follies, and ever more fiercely did Heathcliff pursue his remorseless craving for vengeance. He played Hindley for the possession of Wuthering Heights, and won, and then reduced him to a menial position on what were once his own lands. Hindley's life ebbed away in misery and degradation, and Heathcliff, determined to drain the cup of vengeance to the dregs, turned on his enemy's boy, Hareton, and abased him to a farm-boy's level, as he in his youth had been abased.

But, as the years wore on, a higher Power intervened

Figure 10.

107

"WUTHERING HEIGHTS"

EDGAR LINTON AND CATHY.

in this devil's game. Cathy's baby (Catherine by name), now grown up, chanced the way of her cousin, Hareton, and discovered their relationship. She took pity on the poor brutalised boy and taught him to read and to write. And pity was ever akin to love!

When Heathcliff surprised her in his house one day, he prepared for what he thought would be his culminating act of revenge. She should not leave the house until she —the daughter of one who shrank from him in his degradation—had married the uncouth farm-hand, and in that way he would pay back in kind the slights showered upon him. His ruse answered, and he even gained possession of the title deeds of Thrushcross Grange. But, at the very moment of triumph, he failed completely and ignominiously.

Figure 11.

"WUTHERING HEIGHTS"

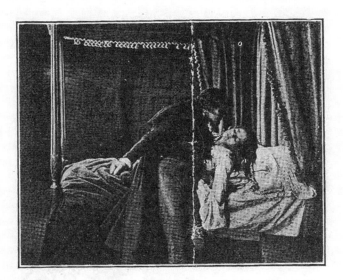

THE DEATH OF CATHY.

When, one day, Catherine presented herself before him in her mother's gown which she had found in the house, Heathcliff—his mind always running on his lost love—leapt forward with outstretched arms, thinking that Cathy had appeared to him. Then, when he realised the truth, he would have beaten Catherine without mercy, had not her screams brought Hareton to the scene, who closed with Heathcliff and hurled him to a safe distance. Catherine clung to her rescuer and looked up to him with an expression of tender love, and that one look came to Heathcliff with the shock of a revelation. Hareton and Catherine did not loathe one another, as he thought, but loved—and with their love all his hate-begotten schemes came tumbling to earth. Like a man awaking from a nightmare, his distracted gaze followed them from the room and, as the door closed behind them, he buried his hands and wept convulsively.

Figure 12.

"WUTHERING HEIGHTS"

"THRUSHCROSS GRANGE."
(Yorkshire Observer Photograph)

And ever afterwards the thought possessed him that he could have loved like Hareton had Cathy but waited for him, and at times it seemed to him that perhaps she was waiting, as indeed she had promised on her death-bed.

And Heathcliff's face, lost its hardness, and became beautified with hope and faith...... And the months drifted by, and one night Cathy appeared to him and he called joyously on her name and kissed the phantom cheek. Then the spirit faded away, and Heathcliff fell dead. For the evil in him had perished utterly, and in all-conquering love—the love of the woman he had now rejoined—he had found the real Power, and the only happiness.

Figure 13.

I have been influenced by the vocabulary of three instructive pieces exploring some fundamental aspects of the aesthetics of British silent cinema. These pieces all foreground a facet of the work of the adaptator of *Wuthering Heights*, Eliot Stannard.

The first piece originates from Charles Barr. In his essay on Stannard, 'Writing Screen Plays' (2002), the word 'tableau' appears in the meliorative phrase 'tableau-style', and replaces the adjective 'pictorial'. This directly echoes the positive lexicon and meliorative views that Andrew Higson developed in his inspiring epic on early British cinema, *Waving the Flag: Constructing a National Cinema in Britain* (1995) as well as in his subsequent discussion of Cecil Hepworth's work, '*Alice in Wonderland* and the Development of the Narrative Film' (2002). The title of Barr's essay itself, 'Writing Screen Plays', is a tribute to the book entitled *Writing Screen Plays* that pioneering British screenwriter Eliot Stannard published in 1920. There, Stannard professes that his favourite artist is the (post) pre-Raphaelite painter, John Collier, whose paintings 'each tell a dramatic story in addition to being masterpieces of artistic composition'.[132]

The second piece that inspired me belongs to Ian W. Macdonald. 'The Silent Screenwriter: The Re-discovered Scripts of Eliot Stannard' (2008) focuses, in great part, on Stannard's interest in painterly composition, within the space of the cinematic frame:

The stage practice of using tableaux had transferred to the cinema screen, and Stannard's scripts usually ended with a tableau-like static grouping (e.g. *Mr. Gilfil's Love Story* Ep.28/Sc.160) and occasionally required tableaux elsewhere where a narrative line came to an apparent end (e.g. *Mr. Gilfil's Love Story* Ep.16/Sc.73), almost always accompanied by an iris closing.

The third and last piece, Christine Gledhill's 'Coda: Hitchcock, *The Manxman* and the Poetics of British Cinema' (in *Reframing British Cinema (1918-1928): Between Restraint and Passion*) was published in 2003. It refers to 'picture portraits and group tableaux' and to '[t]he final curtain tableau'.[133] The vocabulary that Gledhill uses there is of importance since her whole monograph is written in a very precise style. Earlier in the same volume, with the discussions entitled 'Pictorialism and Modernity' and 'Cecil Hepworth: Pictorial Poet of British Cinema', Gledhill exposes the problems of continuity that British 'pictorialism' – the picture seen 'as the centre of pleasurable perception' – could generate when she turns the spotlights on to the debate between advocates and detractors of a 'full continuity system', a debate that was raging in the British film industry and trade press in the early 1920s.[134] The much admired cinematographer George Pearson, in his 1923 'Address to the Faculty of Arts', denigrates the very term of 'continuity', calling it 'that blessed word from America', since he clearly resents being told to 'focus on the production of drama or story' by lesser professionals than him. As Gledhill comments:

Imagining film stories in terms of theatrical stages and pictures posed a challenge to a modernizing film industry seeking the capacity for narrative fiction now established by Hollywood. In particular, pictorial framing and frontal shooting inhibited the scene dissection facilitated by analytic editing, on which depended the illusion of a seamless fictional world that was fast becoming the norm.[135]

Further on, Christine Gledhill argues that the 'coherence of a fictional world', which may not always have been a firm reality in the Pearsonian rationale and practice, was actually ingrained into the Hepworthian conception and production of films, then cogently infers that a '*mosaic*, made up of finely balanced pictorial fragments' should be regarded as a compositional method close – and not opposed – to 'analytic editing'.

*   *   *

Adopting Gledhill's perspective and recognising that the 'tableau-style' and 'montage' approaches are not mutually exclusive – that they can lead up to the same common narrative goal – one can start distinguishing the resemblance between the 'modern' cinema of Hitchcock and the cinema of his British precursors, notably Cecil Hepworth, Graham Cutts and Eliot Stannard. Was it not, after all, the same Stannard who was Hitchcock's regular screenwriter, on his first nine silent films, made between 1925 and 1928? As demonstrated by Kerzoncuf and Barr in *Hitchcock: Lost and Found* (2015), Hitchcock's aesthetical debt is manifest.

Keeping the notion of modern pictorialism in mind, it would be interesting to determine how much narrational information was integrated into the pictorial frames of the 1920 *Wuthering Heights* and conveyed through expressive acting, and how much was dynamically embedded in the unfolding sequences – and fully revealed through editing. In addition to pictorial composition and montage, one can also imagine that some narrational ellipses, distinct from the ellipses deftly inserted by Emily Brontë in her own text, would have been invented by Eliot Stannard (and Albert Victor Bramble) in order to maximise the impact of the adapted text, and be playful with the intertextual knowledge of their contemporary audiences.

What about the rendition of the moralising voice-over that we identified earlier in the *Wuthering Heights* synopsis of Ideal? It sounded like a human voice but, on the cinema screens, probably transmuted into a live musical accompaniment and a series of intertitles, not into a narrator in the flesh interpreting the pictorial frames of *Wuthering Heights*. For at that point in time, as Charles M. Berg indicates in his 1975 article, 'The Human Voice and the Silent Cinema', the film narrator had already become an 'extinct species',[136] which is also the down-to-earth conclusion that, three decades later, Tom Gunning reaches in 'From the Opium Den to the Theatre of Morality'.[137] In that eighth chapter of his *Silent Cinema Reader*, Gunning turns briefly to the celebrated 1979 monograph by Noël Burch, *To the Distant Observer: Form and Meaning in Japanese Cinema*, to uncover the 'cultural reasons why film lecturers would find themselves at odds with the complete narrative development of film (while existing in Japan [with the *benshi*]) through the silent era'; the need for 'a univalent, homogeneous diegetic effect', assumes Burch, and through him, Gunning:

The benshi removed the narrative burden from the images and eradicated even the possibility of the images producing a univalent, homogeneous diegetic effect. In the West, on the other hand, the need for such unity was so strongly felt that it gradually resulted, towards the end of the silent period, in a tendency to do away with titles altogether or nearly so, and 'let the pictures tell the story'.[138]

This ruthless elimination, in the West, of the film narrators – and with them of the prerogative of the audience to narratorial variety and pictorial ambivalence – has been theorised by Kristin Thompson in her seminal essay, 'The Formulation of the Classical Style, 1909-28' (1986). The British films of the period, she argues, while steadily retaining their national specificity, were developing into self-sufficient shows that aimed, increasingly, for the kind of smooth narrative flow soon to be imposed as a norm by Hollywood – namely, the 'classical narrative style'. We can thus safely infer that the moralising voice of the *Wuthering Heights* synopsis was only heard mentally by the readers of the Ideal film programme, prior to their being transformed into 'Ideal/ideal' spectators, on entering the cinema theatres.

Still, the imprint of pictorialism and nineteenth-century melodrama on the *Wuthering Heights* of 1920 endows Bramble and Stannard's creation with a truly dissenting quality,

Figure 14. Frame Enlargement: Heathcliff in the Doorway

in that their film cannot be associated too closely with the classical narrative style. It naturally follows that, as twenty-first century viewers immersed in their memories of (Hollywood) classical movies, we need to find ways of adjusting to the idiosyncratic style of acting that dominates some of *Wuthering Heights'* key emotional scenes, and suffuses this frame enlargement where Heathcliff appears to be transfixed in the doorway. In the late 1910s and early 1920s, such accentuated acting, consisting in pronounced gestures and facial expressions, was not uncommon at all.

Attuning our senses to melodramatic acting can be facilitated by the rationalising of the 'pictorial-theatrical practices' by Christine Gledhill,[139] and the decrypting of early screen acting by Jon Burrows, in his 2002 essay on the filmed adaptation of *Richard III* (1911). There, Burrows gives us some very concrete opportunities to grasp the nature of the 'pictorial style of acting', as well as understand how it actually differs from the art of pantomime, in which it is also profoundly rooted:

> Ben Brewster and Lea Jacobs [in Theatre to Cinema: Stage Pictorialism and the Early Feature Film, 1997] have recently distinguished what they define as a pictorial style of acting, which can be seen to persist in European cinema to at least 1918, from pantomime per se. Whilst pantomime may share a similar gestural repertoire, it is more intensely stylised than the broader tradition of pictorial acting and foregrounds constant performative movement within more stereotypically ritualized dramatic situations. Unlike the pictorial style, which used pronounced gestures to emphasize and illustrate key emotional points of dialogue, pantomime abhorred even the imitation of speaking and listening and placed the focus of attention entirely upon summatory gestures.[140]

This pictorial style of acting is a specificity of early feature films, and is epitomised by the frame enlargement of a Heathcliff, seemingly transfixed in the doorway, as 'he determines to leave Wuthering Heights'.

A bridge does exist between the aesthetic (and ethical) requirements of today's film audiences and the aspirations of the picture-goers who enjoyed the lost *Wuthering Heights*, and were thrilled by these moments of outlandish theatricality that enrich the whole texture of its film narrative. The melodramatic voice, more repressed in today's cinemas, may have become a fainter indicator of the artifice of narration, but the sense of its presence has remained the same. It is akin to the meaning and modus operandi of the Fantastic discourse, since melodrama complements realism as the Fantastic complements the real:

> [...] tied to the conventions of realism, which are absolutely necessary to authenticate and legitimate its emotional and ethical assertions, Melodrama [/the Fantastic] upholds the presence of overriding ethical imperatives in the everyday world.[141]

Chapter 15

Ideal's Synopsis - The 'Ephemera' of the Lively Arts

The very dimension of expressive acting (and emotionally enriched reality) connects with the balladic layer of Emily Brontë's novel and, with the awareness of that deep balladic tonality, the Ideal synopsis sounds less histrionic and unnatural. A striking balladic leitmotif, in this synopsis, consists in the silhouetting of the dark hero, Heathcliff, and in the particular attention given to each of his movements:

> [...] mounting a horse, he disappeared moor-wards.
> [...] Heathcliff kneeling, frantic and maddened, beside her deathbed.
> [...] Heathcliff [...] leapt forward with outstretched arms, thinking that Cathy had appeared to him.

As in a ballad, the sharpness of a depiction made, predominantly, of objective bodily terms amplifies the dark hero's gestures and movements, conveying a sense of foreboding and tragedy that adds an extra dimension to the real. Vachel Lindsay, together with Gilbert Seldes, the author of *The Seven Lively Arts* (1924),[144] was an early theoretician of the 'moving' picture. In his monograph, *The Art of the Moving Picture* (1915), he explains that the 'key words of the stage [are] passion and character' and of 'the photoplay splendor and speed'. Lindsay's words aptly describe the dark hero of the *Wuthering Heights* of 1920 who belonged to two artistic worlds, the world of the theatre and the world of the cinema.[145]

Both theoreticians, Lindsay and Seldes, greatly influenced the chief reviewer of the British *Film Society* (1925-1939), Ivor Montagu, who would soon build on their systems of ideas when commenting on cinema's remarkable ability to accentuate, and even to celebrate movement, through the agency of light and the possibilities of unlimited space.[146] Accordingly, it is Seldes who, less than ten years after *The Art of the Moving Picture*, might have hit on the most compelling formula to synthesise the essence of the 'motion' picture – and of cinematic expression, in general. Cinema's magic formula resides in the 'Ephemera' of the Lively Arts, those arts 'which specifically refer to our moment, which create the image of our lives'. When applied to non-independent, institutionalised cinema, Seldes' concept of 'lively arts' might seem slightly out of focus as the Seventh Art often loses its lightness (and 'Ephemeral' charm) to television, which absorbs more easily the moods, fears and desires of the present times. This being said, while the vertiginous progress of the World Wide Web, digital television and mobile technologies has boosted the availability of films, plays and serials, the chances for them of being seen by live audiences, through wide public (or small family-like) screenings, have not increased proportionally. In the space separating the inanimate screens from the live audiences (or isolated tablet users), it has now become a fundamental mythocritical activity for each spectator, for each viewer to find the energy to re-imagine and even re-invent and open up, time after time, generation after generation, to the many possibilities of live plays envisaged by the *metteurs en scène* and film editors of 'images whose meanings are infinite and ambiguous'.

In the book dedicated to his much-loved predecessor, Yasujiro Ozu, the cinema scholar, documentary and film-maker Kiju Yoshida makes an illuminating comment on the quasi-magical property of the moving images to reflect some *beautiful*, orderly versions of the *chaos* of our lives:

> Ozu-san probably regarded the world as chaotic. He also considered cinematic expression to be chaotic and artificial, thus corresponding to the world. Nevertheless, he dreamed of presenting an orderly world in his films, only within the imaginary universe of light and shadow on the screen.
>
> Therefore, his viewers have to closely stare at the screen, to move back and forth between the orderliness and the chaos, and put up with the images whose meanings are infinite and ambiguous. [...] Ozu-san playfully requested that viewers live uncertain and chaotic lives together with the actors in his films.[147]

The Japanese Ozu and American Seldes' shared postulate on cinema's identity – an art re-enacting the Ephemera of human lives – reinforces the idea that a well-crafted period piece like Ideal's adaptation of *Wuthering Heights* represents a particular vision of Emily Brontë's novel (that of the late 1910s) rather than the novel *per se*. Ideal's adaptation eerily reflects the shadows of its screenwriter Stannard and director Bramble; it also reveals the outline of its past audiences and anticipates, to some extent, the audiences to come.

The curse of this silent picture (and of so many contributions to silent cinema) is the silver nitrate emulsion used for the master negative and the prints: a chilly atmosphere of around 4 degrees Celsius is required to save them from spontaneously catching fire. Missing the opportunity of being copied back on to a 35-mm film (then reformatted into a Blu-ray disc, for instance), some movies of that period have vanished altogether.

The *Wuthering Heights* of 1920 might have been amongst the nitrate films that were seen as mere commodities and recycled for their high silver content. It might, on the contrary, have been considered a valuable corporate asset by *Gaumont-British* – who bought Ideal in 1927 – but suffered careless handling (or inexpert storage) and, eventually, perished.

B

Recomposition of *Wuthering Heights* through Stannard, Bramble and Some Shared (or Related) Pictures: An Insight into the Hermeneutic of *'Incursion'*

Chapter 16

A Modern (Silent) Motion Picture

I deal's synopsis, and all the written and pictorial evidence available, suggests a rich cinematography that abounds in cathartic and picturesque scenes, bringing into play the spectators' foreknowledge of the novel, and greatly stimulating their imaginations. The extant script of *Mr. Gilfil's Love Story*[146] – the picture that precedes *Wuthering Heights* and springs from the same artistic collaboration (Stannard and Bramble) – could serve as a basis to infer safely how the film-makers conceived the key scenes of their next adaptation: some tableaux possessing their own internal logic, which merges into the external logic of the *Wuthering Heights* moving picture itself:

> They [these groupings of actors] effectively give us editing within the shots as a workable alternative to editing between them.[147]

Based on Stannard's re-discovered scripts and other writings – such as 'The Art of Kinematography No.1 – Symbolism', published in the *Kinematograph and Lantern Weekly* on 23rd May, 1918 – as well as on Gerald Turvey's essay 'Enter the Intellectuals: Eliot Stannard, Harold Weston and the Discourse on Cinema and Art' (2003),[148] film scholar Ian Macdonald further argues that Stannard was consciously achieving a modern cinematic aesthetics. He was seemingly able to 'pre-visualise' the film in its key narrative elements and mastered an elaborate syntax of dissolve and superimposed fades, cross-cuttings and close-ups. Moreover, Stannard had an excellent sense of tempo and feel for suspense (or 'time-anxiety'), and was constantly focusing on continuity and thematic unity:

> Stannard's symbolism did not occur merely as occasional montages or flashes of static compositions; he thought of the whole of kinematography as 'The Art of Symbolism', meaning 'the manufacturing of a series of rapid impressions which will, in a given time, convey this or that story to the audience.'[149]

At the time of the 1920 *Wuthering Heights*, only two things seemed to have been missing from the British silent pictures. The first one was the ordinary picture-goers' legitimacy as film commentators. When *Wuthering Heights* was released (25th Oct, 1920), a film spectator who, unlike Mr Jonas Bradley, 'advisor on *The Brontë Centenary at Haworth* (1917)', had not been accredited in some form of fashion by the Brontë Society (and had almost certainly not participated in the making of the movie), could not dream of having their opinion relayed in the national or regional press. This typical, anonymous film spectator could not presume either to have their opinion recorded totally independently from Ideal, the production company. For, if the film exhibitors were testing quite liberally Ideal's dedicated feed-back cards on their patrons after the shows, Ideal would understandably only select the best ones for publication in the trade magazines. Further, such fan magazines as *Pictures and the Picturegoer* – or the London-based *Picture Show* and *Girls' Cinema* – functioned mainly as vehicles for promoting the pictures and star performers, and had not yet started to treat their readership as an audience worth talking to after the shows.

Thus, leaving aside the impersonal and succinct comment issued by the Brontë Society in their 27th annual report (19th Feb, 1921) – '[d]uring the year a good representation of Emily Brontë's *Wuthering Heights* has appeared for the first time on the cinema screen, and has met with general approval' – the audience response to the 1920 *Wuthering Heights* consists in shadows of isolated feed-back cards, mainly designed by Ideal as potential publicity ploys for selling their forthcoming shows to the exhibitors even better. One of those ghost cards actually materialised, and is reproduced on the next page. It was hand-written by the manager of the Pavilion cinema, Uddingston (Scotland), forwarded to Ideal who could only approve of it, then published in the *Kinematograph Weekly* nine days before the release of the bland report issued by the Brontë Society. In the absence of a cluster of feed-back cards, this exceptional artefact that reveals the (mediated) thoughts of an ecstatic spectator comes short of reflecting the general appreciation of the Scottish public for the Bramble-Stannard adaptation but cannot help aptly describing the 'Heavenly', *beautiful* side of Emily Brontë's novel:

'I thought I was in Heaven tonight' these were the words of one of my patrons as she left my place after having seen the above.

The second element missing from the British silents was at the time, of course, 'natural sound' as argued with a certain nostalgia by Alfred Hitchcock in his interview with François Truffaut, back in the late 1960s:

Well, the silent pictures were the purest form of cinema; the only thing they lacked was the sound of people talking and the noises. But this slight imperfection did not warrant the major changes that sound brought in. In other words, since all that was missing was simply natural sound, there was no need to go for the other extreme and completely abandon the technique of the pure motion picture, the way they did when the sound came in.[150]

This literal 'abandon of the technique of the *pure* motion picture' is of paramount importance in the thwarted destinies of both Albert V. Bramble and Eliot Stannard.

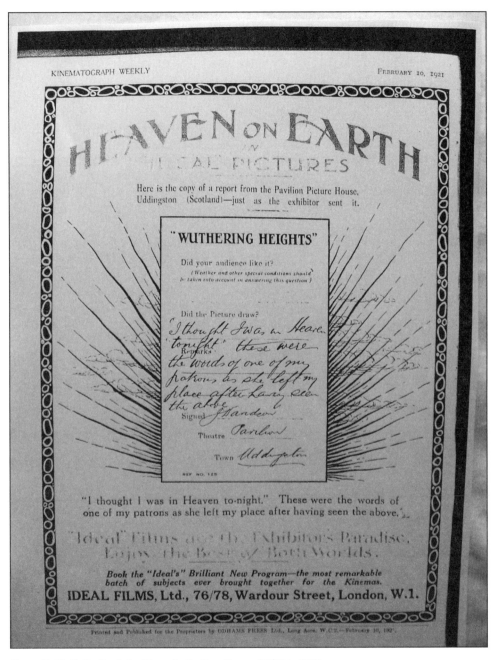

Figure 15. Ideal's Feedback Card from the Pavilion Cinema (10th Feb, 1921)

Chapter 17

Recomposition of *Wuthering Heights* through the Elusive Director Albert Victor Bramble

The scriptwriter Eliot Stannard comes back to life through his theoretical writings and his extant scripts – in particular, *Mr. Gilfil's Love Story* (1920). The director Albert Victor Bramble, despite the apparition of his striking figure on two outdoor photographs taken during the shoot of *Wuthering Heights*, does not materialise as easily as his fairly visible collaborator. Their *Wuthering Heights* production will remain for Bramble, until the end of his cinematic career, the best picture he made as a film director, and the two snapshots of him directing on location reflect his total involvement and reveal the spirit of cooperation that demanded the shoot on the moors. As highlighted by *The Yorkshire Observer*, in its article entitled '*Wuthering Heights* in a Film Version – Mummers on the Moors – The Difficulties of Cinema Play Production' (Wednesday, 5th May, 1920), the success of Bramble's shoot on the moors depended not only on his and his crew's sheer physical efforts (when it came to carrying the equipment around, for instance) but also on the ability, for each cast member, to endure the harshness of the heathland while giving expression and nuance to their performances.

The cameraman, side by side with Bramble, is at the centre of the two 'on-location' photographs that are presented on the following pages. At that time, Ideal only counted four 'camera experts'.[151] Looking closely at these two images, it becomes clear that the ever smartly dressed cinematographer Claude McDonnell – who, in 1921, would operate his camera during an expedition led by Ernest Shackleton, the Antarctic explorer, then assist the young Alfred Hitchcock on his first films at Gainsborough – was the one chosen by Albert V. Bramble for his *Wuthering Heights*.

In the first still, Bramble, donning a felt hat and a long coat, stands right beside McDonnell, who has balanced himself on a simple wooden chair to work his camera. Under the watchful eye of Bramble's assistant, Miss Murray (who wears a warm coat too), and a couple of technicians holding the white, window-like panels that reflect the sunlight on to the filming area, they are shooting the scene, non-existent in the book, where Cathy is rescued by Edgar: she has been rambling on the moors with a boyish Heathcliff, and sprained her ankle.

In the second still, Bramble and McDonnell, both sporting tweed suits and high boots, are wading in a rocky stream and carrying, between them, the tripod and the camera box in preparation for, or possibly just after, the scene involving Edgar, young Heathcliff and Cathy, at a spot known as the 'Brontë Waterfalls' – that is, according to the already cited article from *The Yorkshire Observer*.

Figure 16. Outdoor Shooting: The Cameraman and A.V. Bramble

Like many cinéastes of his generation, Albert Victor Bramble spent his formative years in the theatre where he mixed with some highly esteemed professionals, including the stage producers and actors, Oscar Asche and Fred Terry. He communicates quite a few valuable facts and details in an unidentified press biography that appears to date back to his comeback role in *Outcast of the Islands* (1951), when he was seventy years old.

This press biography is unusually long at five pages, written in the first person and, with its erratic punctuation, sounds like the transcript of an interview.[152] The personal style could be deceptive, though, since Bramble's degree of participation – and his latitude for revision – cannot be evaluated until some of his own papers, or an authenticated interview of his, re-surface. Still, the very first sentence that reads 'I set out in life to become a *Black & White Artist* [emphasis added]' is as much striking for its affirmation of a very early vocation as for its unusual metonym, '*Black & White Artist*', which characterises the film-maker as a photographer – or a painter – playing with light and shadows in order to create contrasts and movement.

Figure 17. Outdoor Shooting: A.V. Bramble and His Cameraman in the Stream

Albert V. Bramble was relatively mature when, in 1914, at the age of thirty-four, his old connection with Maurice Elvey at Fred Terry's helped to kick-start his film career. In his friend's first picture, *Black-eyed Susan*, a nautical melodrama adapted from a popular comic play and commissioned by the British and Colonial Kinematograph Company (B&C), Bramble plays the part of the Pirate Hatchet. Then, in 1915, he somehow manages to keep working as an actor on all of Elvey's films (but one) while directing his own full-length picture at B&C, *Hearts that are Human*. This directorial breakthrough was soon to be followed, in 1916, by *Jimmy*, the first feature film resulting from a collaboration with Eliot Stannard, who 'wrote the scenarios for nearly all [his] silent films'. Stannard's script for *Jimmy* is adapted from the novel (1903) written by his own mother, John Strange Winter – her 'nom de plume' – and the film, led by the close-knitted Stannard-Bramble duet who had already worked together at production level (on two shorts at least), sees Bramble in the main part as the old banker, John Denbeigh, whose rebellious son is transformed beyond recognition by his experience of the war. This role fitted Bramble like a glove as, says an undentified third-person publicity biography that could emanate from Master Films, he was 'first and foremost a character actor'. This fine actor-director declares himself that 'of all the film parts [he] has played none has given [him] more satisfaction and pleasure', despite the burning topicality of the story.[153]

The film career of Maurice Elvey, who had become a fixture of the British film industry in the mid-1910s, lasted until the late 1950s. Albert V. Bramble's own film career did not. One lead for his relative lack of longevity as a film director was his unshakeable devotion to the stage, where he kept returning at regular intervals, in-between his film shoots. Yet another lead for Bramble's (and, in the same breath, Stannard's) progressive erasure from the film credits points to some difficulties of acclimatisation during the particularly tough times of the transition to sound. This 'sound' revolution – which can still be fully appreciated on viewing the silent and sound versions of Hitchcock's *Blackmail* (1929)[154] – accelerated the change from a cooperative and flexible practice, where the director, producer and screenwriter could occupy some interchangeable roles, to a much more rigid and (big) studio-based process:

> The impression in the earlier period is of a different system of relationships; flatter, less structured and with less well-defined roles, and where the writer's word was perhaps more of an instruction.[155]

There was this long ten-year period, from 1923 to 1933, when Bramble appeared to have been cherry-picking his artistic projects. He enjoyed being an actor again (on stage or in the film studios) and, in 1923-1924, took an active directorial part in two war reconstruction films, *Armageddon* and *Zeebrugge*, put together by the young and energetic Harry Bruce Woolfe for his own film company, British Instructional Films. Then, in 1928, Bramble would team up with the dashing Anthony Asquith, on *Shooting Stars*, for that same independent production company led by Woolfe, while keeping alive his professional friendship with Stannard and making in the same year, *Chick*, for British Lion Film Corporation. Most importantly, it transpires that, back in 1930, he travelled to the UFA Studios in Berlin to do what would now be called 'film dubbing':

> I got my first insight into the mysteries of a really well equipped sound studio when, in 1930, I went with a band of players to the UFA Studios, Berlin, to put English speech to German-made pictures.

His varied output was steady, and lasted nearly two decades. It stopped abruptly in 1933 after three talkies now lost and practically forgotten, *A Lucky Sweep* (1932), *Veteran of Waterloo* (1933) and *Mrs Dane's Defence* (1933), which he made for National Talkies:

> Alas, this [*The Man who Changed his Name*, in 1928] was to be my last silent picture; even before I had finished it, the 'Talkie' had arrived and with it chaos and a state of upheaval the like of which, I imagine, no other industry has ever experienced.[156]

In retrospect, it is poignant to see how Albert Victor Bramble fell into quick anonymity, a few years only after the silent period had died away. Like Cecil Hepworth who, for an intense nine-month period in 1918, involved him in his productions and became his high-profile mentor, he apprehended film-making not so much as a standardised and self-centred enterprise, but as a collective and artistic venture. Eventually, both Hepworth and Bramble's directorial careers suffered, and came to a standstill. Here is another excerpt from Bramble's first-person biography, which lays bare his close connection with Hepworth, and reveals a great deal of his personality:

> My sojourn with Hepworth lasted about nine months, during the greater part of which time, it was my privilege to act as his assistant; and from him I learned much about the making of pictures. When, on being invited by Master Films to direct for them I left him, it was with great reluctance that I did so.

What line of work did Bramble pursue, if any, during his lengthy pre- and post-war absence from the visible side of the cinema industry? Somehow, he was not completely forgotten, and must have enjoyed a fleeting moment of recognition when, in April 1950, his peers made him an honorary member of the *British Film Academy*. He subsequently effected his comeback on-screen with his part in *Outcast of the Islands* (1951) directed by Carol Reed, and starring Trevor Howard as the wayward Peter Willems.[157] In this adaptation of Joseph Conrad's second novel, where the outsiders are by far the most memorable characters, Bramble played his last role, the old and blind pirate chief, Badavi, prefiguration of Willems as an outcast.

With a talent for acting and directing actors, a profound interest in artists and painting and no prejudice whatsoever against new cinematic methods, Bramble seems to have been cut out to be a *metteur en scène*. His qualities d'*homme de scène* et de *metteur en scène* are underlined in this third-person press biography, which probably stems from Master Films:

> He is a great upholder of the British films but at the same time, is a great admirer of American productions and believes and does not mind admitting it, that unless we compete very seriously with them and try and do films as good as the Americans by using their methods, we shall never do anything great in this country.
>
> He is a great lover of the beautiful, will have artistic scenes and settings and is a great stickler for details.

During his time at British and Colonial (B&C) (1914-1917), British Lion Film Co. (1919 and 1928) and Ideal (1920-1922), his name was always associated with that of Eliot Stannard. Stannard even wrote the script of Bramble's first shorts, *The Blind Man of Verdun* (1914?) and *The Boy and the Cheese* (1914). Still, the scarcity of extant footage from B&C, British Lion Film Co. and Ideal would impact profoundly on the visibility of

Bramble's directorial output, and diminish Bramble's importance in his partnership with Stannard. Additionally, the complete absence, so far, of personal or published material originating from his own hand obscures further Bramble's cinematic output, and means that there is yet another hindrance to scholarly re-appraisal.

This archival void made Bramble an easy target in two derogatory statements springing from Rachael Low and featuring in her already cited monograph, *The History of the British Film 1919-1929*.[158] The first one arbitrarily assigns Elvey, Pearson and Bramble to a sub-category of directors, while turning the *British Association of Film Directors* (to which they belong) into a second-rate organisation. Low reckons that it is by chance, and not out of competence, that the Association tackled such a fundamental issue as the over-representation, on the British cinema screens, of the Hollywood-made American films:

> Another association of British film producers was also formed in early 1922 by a group which included Manning Haynes, Fred Paul, Will Kellino, Donald Crisp, Hugh Croise, A.V. Bramble, F. Martin Thornton and the moving spirit and President, Percy Nash. Nash was followed by George W. Pearson and later Maurice Elvey, but few of the better directors seem to have supported it. It was called the British Association of Film Directors, and although its original aims were more those of a professional guild than of an industrial association, it later took some part in the quota discussions simply because there was no more suitable organization to represent the producers.

In her second statement, Low's judgement on *Wuthering Heights* – doubled up by her depreciation of the entirety of Bramble's filmography – comes across as hasty and caricatural since no careful contextual study or visual/textual leads are shown to substantiate what she writes:

> Bramble did not live up to the review of his first film [*A Non-conformist Parson* (1919)] […] He himself seemed in his old age to feel that his best film was the Ideal version of *Wuthering Heights* (Trade Show July 1920) with Milton Rosmer, but there is every indication that this, like the rest of this make, was an old-fashioned production of very poor quality.

Low's hasty assessment of Bramble's output cannot conceal a valuable archival clue: Bramble's positive appraisal of the very picture she belittles, 'He himself seemed in his old age to feel that *his best film* was the Ideal version of *Wuthering Heights*' [Emphasis added] – which stands out conspicuously in the text of her chapter entitled 'Producers in the Early Twenties'.[159] Bramble's excellent opinion of the film is even iterated slightly differently in a later chapter entitled 'Techniques of Film Production: Costs': 'and according to A.V. Bramble his own *Wuthering Heights, a 1920 film of which in his old age he remained extremely proud*, had cost £6,000'.[160] [Emphasis added]

Low might have hit on a diary or some kind of memoir. She could also have been quoting from an interview. In any case, the comment embedded in Low's prose does not appear to have been passed lightly by Bramble: this sounds too much like his own lost voice to be lacking in authenticity.

The loss of most of Albert V. Bramble's filmic and written work also exposes him to some contradictory innuendoes, and even inadequate criticisms. In the chapter dedicated to *Shooting Stars* (1928) in *The Cinema of Britain and Ireland* (2005),[161] the film historian Luke McKernan points out that, in the credits, the acclaimed picture is presented as being 'by' (the then inexperienced) Anthony Asquith but 'directed by' A.V. Bramble, which downplays Bramble's contribution while making it unmissable. Still, McKernan carries on, and utilises Bramble's name in the plural form with the derogatory expression, 'a host of A.V. Brambles', in order to encapsulate the amateurish nature of cinema practices in Bramble's generation of film-makers – those 'modestly talented journeymen':

> Bramble was a stolid, efficient director of the old school in British film, which in creative terms meant very little. His past credits as a director included *The Laughing Cavalier* (1917), *Wuthering Heights* (1920), *The Card* (1922) and co-director of Harry Bruce Woolfe's *Zeebrugge* (1924). None of Bramble's surviving work indicates any filmic gift, and the relatively long career of such a journeyman talent indicates the impoverished nature of British film in the early to mid-1920s. […]

> There has been sufficient critical argument in recent years to overturn the traditional view of British silent cinema as being characterised by low achievement, but it is unquestionably true that, at the time, all but its most blinkered practitioners viewed British film in a highly critical light. It was a cinema manned by a host of A.V. Brambles: earnest, modestly talented journeymen, who displayed a profound lack of understanding of the medium in which they were working. Directors of minimal vision were allied to technical staff and studio equipment of only the most basic competence, scripts that could not imagine the leap from the printed word to the screen, and performers caught between the mannerisms of the theatre and a desperate aping of the stylings of American film.

Further, McKernan opportunely contrasts him with the likes of Anthony Asquith, Alfred Hitchcock and even Maurice Elvey while conceding, perhaps half-heartedly, that 'the *old hand Bramble* was brought in and given the director's credit, principally for *his knowledge in the handling of actors*' and that 'the film is notable for *a realism in performance which alone marks it out as something new in British film* – and *something for which A.V. Bramble should probably receive more credit than he has previously been allowed*'.[162] [Emphasis added]

Interestingly enough, the remarkable input of Bramble in *Shooting Stars* was acknowledged by Anthony Asquith himself in a letter, which is not dated but was clearly written soon after *Shooting Stars*, and addressed to Adrian Brunel. This letter is held at the British Film Institute (BFI) in the Box 170 of the Brunel papers:

> Bramble has been extraordinarily good and it would be very hard if he did not get credit.

Lack of footage and professional writings has led to the disparagement of Bramble's film career and a misrepresentation of the moving picture he was most proud of, *Wuthering Heights*.

*   *   *

After having only been mentioned briefly in the chapter, 'A Modern (Silent) Motion Picture', Stannard, the wonderful screenwriter who lent his talent to Hitchcock on his first nine films (produced at Gainsborough and British International Pictures) now needs to be summoned again. Here is what Charles Barr writes about him in his article entitled 'Writing Screen Plays', part (again) of the superb collection, *Young and Innocent?* (2002):

> On Hitchcock's first nine films as director, released between 1926 and 1929, the single credited screenwriter other than, on two occasions, Hitchcock himself, is [the same] Eliot Stannard; he has solo credit on seven out of the nine, and a shared credit on *Champagne*, and his name is absent only from *The Ring*, to which he is known, however, to have made some kind of contribution as well.[163]

In many ways, Stannard emerges as the key character for lifting from oblivion *Wuthering Heights*, the film he wrote for Bramble, and allowing his old comrade, with whom he worked on so many other productions, to find his rightful place in the field of early British cinema.

Mr. Gilfil's Love Story (1920) and another shared production, *The Bachelors' Club* (1921), attest to a congruence between the writer's screen idea (Stannard's) and the director's screen realisation (Bramble's). Even more crucially perhaps, a later picture made by the Stannard-Hitchcock team, *The Manxman* (1929), bears some striking thematic and aesthetic similarities to Stannard-Bramble's *Wuthering Heights*; it showcases a pictorial cinematography of framed landscape, portraits and group tableaux that generates a strong emotional response in the viewers.

Chapter 18

Incursion into *Wuthering Heights* through the Bramble-Stannard Partnership and the Parallels between *Wuthering Heights* and *Mr. Gilfil's Love Story*

A classic hierarchy with the director (or the writer) at the top of the film-making pyramid does not necessarily correspond to the organisational realities of *Wuthering Heights*, if one considers for an instant the small team of close collaborators hand-picked for this venture. We could well imagine Stannard or even the cameraman, Claude McDonnell, participating in the scouting for locations and, later on, in the selection of film stocks.

Conversely, Albert V. Bramble, in view of his deep involvement in a picture he would later consider to be his best, may well have taken more initiative than usual during the key phases of literary adaptation and picturisation (which were paving the way for his filming), and injected his own ideas into Stannard's script.

It is Ian Macdonald who puts forward some archival tip-offs. Some of these tip-offs are situated in this part of 'The Silent Screenwriter' (2008) where he decrypts the script of *Mr. Gilfil's Love Story*, while others are traceable to his preceding article, 'The Struggle for the Silents' (2007), which contrasts the writer's draft (Stannard's) and the shooting script (Bramble's) of *The Bachelors' Club*.

The Bachelors' Club is the focal point of 'The Struggle for the Silents'; it is a cultural translation of Israel Zangwill's witty series of stories published in 1891. The written material that survived consists of two complementary scripts that are rarely found together: Stannard's draft (Format A) and Bramble's shooting script (Format B). This is quite extraordinary. In most cases, the shooting script (or continuity) never materialises. It may be that it is jealously kept in some bookcase and remains unnoticed, or is regarded as valueless and lost. Comparing these two versions, Macdonald deduces that 'the writer presented a complete visual concept of the film from the initial script' and postulates that the cinematographic industry of the time (1918-1933) 'had not yet rationalised the growing dominance of the director into something more than the person who *realised* what the writer had *picturised*' [emphasis added]. He also adds warily that 'early screenwriting practice in Britain' still required more research, and cautions against the distorting effects of the 'auteur' theory on early British film-making.

In the first paragraphs of 'A Modern (Silent) Motion Picture', it is the extant script of *Mr. Gilfil's Love Story* that lends some clues as to the modernity and efficiency of Stannard's screenwriting. From the perspective of Bramble's well-loved *Wuthering Heights*, this script also signals a fine complementarity between Stannard's and Bramble's work and announces the expertise necessary to tackle – after only a short breathing space – a novel

Figure 18. 'Ideal's' *Mr. Gilfil's Love Story*

that, like George Eliot's fiction, has a strong melodramatic potential. However, prior to focusing on Macdonald's words, it would be profitable to follow George Eliot's (alias Mary Anne Evans) train of thought when she envisioned the central tale of her *Scenes of Clerical Life* (1858). The short cut is provided by Laura Mooneyham White in her audacious article 'Melodramatic Transformation: George Eliot and the Refashioning of *Mansfield Park*' (2003):

> *Mr. Gilfil's Love Story*, after all, takes for its frame story the fate of the eponymous Mr. Gilfil, who gathers Caterina into a safe marriage after her own 'love story' (the subject of the central narrative) has ended in death for her lover and emotional ruin for herself. Caterina becomes inscribed within the story of a virtuous clergyman – boxed, as a morally disturbing exemplar, within the life history of a kind but dull man. [...] Both titles, then, *Mr. Gilfil's Love Story* and the volume's *Scenes of Clerical Life*, are misleading. Emotional, rather than clerical, life is key to all three stories, and it is Caterina's love story, not Mr. Gilfil's, that forms the second story's centre.

As compellingly demonstrated by Mooneyham, Eliot was recasting an older story with an emphatic preference for the Romance over the Novel, and the *sublime* over the *beautiful*, thus upgrading rather than discarding emotion and passion:

> *Mr. Gilfil's Love Story* recasts the narrative of *Mansfield Park* [1814] in ways that reveal Eliot's project of correcting Austen's deficiencies by 'rais[ing] the passions into sympathy with heroic troubles'. To do so, Eliot must retell *Mansfield Park* in what she clearly felt was a more emotionally realistic way. Ironically, Eliot's attempt to remedy what she sees as Austen's deficient realism leads her directly into the conventions of melodrama, and led her to create a story almost unreadable today because of its extravagant affective appeals.
>
> [...] That Eliot recaptures some of the 'egregious affectivity' that marked the work of authors such as Wollstonecraft and Radcliffe shows a reaching back for emotional rather than ironic authority. Austen's role for Eliot, her use, is to furnish a plot, the capital, of *Mr. Gilfil's Love Story*, and to submit to its reauthorization by means of emotional investment.

This 'emotional investment', which connects aesthetically and politically with what Emily Brontë achieved in *Wuthering Heights*, was acknowledged by Stannard and Bramble. They kept close to the dynamic structuring of Eliot's novel while Ideal, the production company, decided on the global adaptative strategy – and probably selected themselves Eliot's title, *Mr. Gilfil's Love Story*, to inaugurate 'the Big Seven', their high-profile 'Batch of Trade Shows'.

The heroine, Caterina Sarti (a role tailor-made for the very young actress Mary Odette) and not Mr. Gilfil (Henderson Bland!), was featuring at the heart of the picture. The film-makers, Bramble and Stannard, aimed, with this essential thematic choice, for an 'authentic' translation of the source material provided by George Eliot. Their translators' approach, quite bold, departed from the more stereotyped views on *Scenes of Clerical Life*, epitomised in the *Bioscope* review of 22nd Apr, 1920:

> If Gilfil is the protagonist (as he really ought to be), he should be given far greater prominence in the earlier episodes; in the film version, he is quite a minor character throughout.[164]

Macdonald explains, in 'The Silent Screenwriter' (2008), that the writer's draft of *Mr. Gilfil's Love Story* was the script used by Bramble. It has 'pencilled notes referring to the footage of takes on each scene', and there are also 'cuts and new shots noted in pencil':

> This was director A.V. Bramble's copy, of course, and it surely reflects the director's intentions for the film at time of shooting, whether or not a continuity [or shooting script] was also used to ease shooting management.

The script of *Mr. Gilfil's Love Story* – and all the surviving scripts springing from the cooperation Bramble-Stannard – belongs to Albert Victor Bramble's newly uncovered set of papers. These scripts were acquired by the collector and scholar Fred Lake who quickly offered them to Ian Macdonald for interpretation. They reveal Bramble's profound interest in the whole process of cultural translation, inclusive of writing and pre-visualising. While the hard evidence of changes to the script of *Mr. Gilfil's Love Story* only corresponds to traces of directorial amendments, the implication is that Bramble's commitment was never flinching as the screen idea was being developed, and that his involvement, on such an inspiring project as *Wuthering Heights*, could have occasioned an overlapping of function highly beneficial to the quality of the translation process.

Again, in the absence of 'pencilled notes referring to the footage of takes on each scene' – or to 'cuts and new shots noted in pencil' – no conclusion can be reached about the conceptual phase of *Wuthering Heights*; however, neither the film reviews, nor the journalistic, academic or biographical material, nor the photographic artefacts, nor the evidence linked to Ideal's promotional campaign, indicate that the contribution of Bramble was negligible at any stage of the production. While Stannard probably deserves all the credit for the writing, it only seems fair not to burden him, the 'author' of the script, with the title of 'auteur' – when he himself lectured on the trinity cameraman-director-screenwriter, in his speech to the Kine-Cameramen's Society, on 17th Feb, 1921; rather, his confrère Bramble should be allowed to have shared a vision, and given some considerable impetus to the film project of *Wuthering Heights*.

Chapter 19

Hitchcock's Hidden Collaborator, the Screenwriter Eliot Stannard

If ever a print of the lost film were to re-surface in the archive of a remote museum, as the copy of a mostly complete version of Fritz Lang's *Metropolis* did at the Museo del Cine in Buenos Aires (Argentina) in 2008, Albert Victor Bramble would at long last come out from the shadows, the way he would have dreamt it, with his favourite production.

Eliot Stannard himself is somehow resuscitated thanks to the scholarly interest of Charles Barr that materialised into *English Hitchcock* (1999) – and 'Writing Screen Plays: Stannard and Hitchcock' (2002), which triggered Michael Eaton's article, 'The Man Who wasn't There', widely circulated in *Sight and Sound* in 2005 – as well as, much more recently, into the culminating piece, *Hitchcock and Early Filmmakers*, in *The Companion to Alfred Hitchcock* (2011). Further, Ian Macdonald's already cited articles on screenwriting in the silent British era and his archive-based stunning monograph, *The Poetics of Screenwriting* (2013), have kept Stannard at the centre of discussions, and re-opened the memories to the work of Albert V. Bramble, which means that there might be a glimmer of hope too for the elusive director of the lost *Wuthering Heights*.

In the 'Notes' appended to 'Writing Screen Plays', the distinguished British screenwriter Sidney Gilliat, co-writer with his associate, Frank Launder, of the script of *The Lady Vanishes* (1938) – and adaptator of Daphne Du Maurier's *Jamaica Inn*, also directed by Hitchcock around the same time – wittily attests to the versatile nature of Stannard's work at British International Pictures (BIP). After a string of five promising films with Hitchcock, at Gainsborough, and a few other well-fulfilled engagements, Stannard came to BIP, at the instigation of the young director, who could not afford to lose his writer when moving studios.[165] There, back in 1928, while training in the scenario department, Gilliat met the man who can still get the full attention of a professional scenarist like Michael Eaton, nowadays, with his 'Lesson Six – Writing Screen Plays', pitched, nearly a century ago, for *Cinema – Practical Course in Cinema Acting in Ten Complete Lessons* (1920).[166] Interviewed for John Boorman and Walter Donohoe's book, *Projections 2* (1993), Gilliat came up with this vivid recollection of Stannard at BIP:

> The only resident British writer I can remember was Eliot Stannard, a great character. He seemed to be writing or rewriting everything. If something went wrong on a picture, Stannard was called up – like Shakespeare would have been – and asked to come in and pep up the scene a bit.[167]

Stannard also contributed actively to cinema with critical essays, from 'The Use of Symbols in Scenarios', published as early as July 1917 in the *Kinematograph and Lantern Weekly*, to his subsequent series of five weekly articles on 'The Art of Kinematography', which appeared in the same trade paper in May-June 1918,[168] and even gave a lecture at the *Kine-Cameramen's Society* in February 1921. If these texts do not give us any direct insights into his long collaborative aventure with Bramble, or into the specificities of his aesthetical involvement in *Wuthering Heights*, they all offer some valuable generic clues. In *Writing Screen Plays* (1920), his summative critical piece, he expresses a strong liking for the pre-Raphaelite artist John Collier whose hyper-real paintings, loaded with dramatic intensity, have the crispness of modern photography and the compositional balance of a Renaissance design:

> It is the Scenario-writer's business to visualise each scene he writes as 'a picture'. The Hon. John Collier's paintings are nearly all immobile subjects which might well have formed episodes in a screen play. 'The Cheat', 'Sentence of Death', 'The Confession' and 'Eve' each tell a dramatic story in addition to being masterpieces of artistic composition. If the Scenario-writer can be equally dramatic by careful attention to composition, grouping and detail, his play should hold and grip his audience until the very end.[169]

This admiration for Collier's works bespeaks Stannard's sophistication and aptness to translate on to the screen not only the contrasting qualities of the dwellings (Thrushcross Grange and Wuthering Heights) but also those of the *beautiful/sublime* landscapes and couples of lovers created by Emily Brontë. Thematically and aesthetically, there are even some undeniable parallels to be found between the incomplete jigsaw-puzzle of Stannard and Bramble's *Wuthering Heights* and the intact picturesqueness of Stannard and Hitchcock's *The Manxman*. Back in 1928-1929, this beautiful movie, for which there is no extant script, marked another high point in Stannard's rich filmography.

Unfortunately, it also signified a slow dilution, for this fine screenwriter, of his artistic freedom. The birth of the 'Talkie' and, in the words of Bramble, 'with it chaos and a state of upheaval', pushed him slowly but surely out of business. So did the abrupt end of his collaboration with Hitchcock, who had squeezed all he could out of him and would work exclusively with his unsentimental wife, Alma Reville, thereafter, or, at least, until Charles Bennett was requested to come onstage for the *Man Who Knew Too Much* (1934).

Two essays, Christine Gledhill's 'Coda: Hitchcock, *The Manxman* and the Poetics of British Cinema' (2003) and Mary Hammond's 'Hitchcock and *The Manxman*: A Victorian Bestseller on the Silent Screen' (2008), have supported the case of a connective vision between *Wuthering Heights* (1920) and *The Manxman* (1929). These texts do so by contextualising *The Manxman*'s inherited 'pictorial-theatrical' style, and exposing its modern cinematic impact and progressive moral values.

Further, Mary Hammond's probe into the adaptation history of Hall Caine's novel, which was 'a critical success – even outranking Hardy's *Tess of the D'Urbervilles* in some quarters' – and her inventory of the specifics of its different cultural translations help in identifying the challenge that Stannard had to face as a screenwriter and film adaptator. His considerable interpretative talent becomes strikingly clearer once the term 'film adaptator' is superimposed on to the term 'music composer' in this excerpt taken from George Steiner's 'Topologies of Culture':

[The film adaptator's] initial thrust in the significance of the verbal sign system is followed by interpretative appropriation, a 'transfer into' the [filmic] matrix and, finally, the establishment of a new whole which neither devalues nor eclipses its linguistic source. The test of critical intelligence, of psychological responsiveness to which the [film adaptator] submits himself when choosing and setting his [script] is at all points concordant with that of the translator. In both cases we ask: 'has he understood the argument, the emotional tone, the formal particularities, the historical conventions, the potential ambiguities in the original? Has he found a medium in which to represent fully and to elucidate these elements?'[170]

Chapter 20

Positioning of *Wuthering Heights* (1920) into the Incipient Wave of Poetic Realism Incarnated by Hitchcock-Stannard's *The Manxman* (1929)

The *Manxman* is readily available on DVD as part of *The Early Hitchcock Collection*, and one cannot help noticing, especially after the archival study of the lost *Wuthering Heights*, a striking similarity between its plot and the storyline of the first volume of Emily Brontë's novel. On the second and third title cards, since *The Manxman* is a silent film, the socially antagonistic pair, Pete (cf Heathcliff) and Philip (cf Edgar), are introduced as childhood friends:

> Philip Christian, a rising young lawyer, and Pete Quilliam, the fisherman, met as boys and grew up as brothers. [2nd title card]

> Still the staunchest friends, they fought side by side for the cause of the lowly fisherfolks. [3rd title card]

It is not stated in the film but, owing to their complicity (and the intensity of their feelings), Pete, Philip **and** Kate must have known each other from childhood. The two men's stifled rivalry is revealed though their growing attraction to Kate (cf Cathy), on the subject of whom they keep quiet, that is until Pete, blind to his closest friends' feelings, asks Philip to propose marriage on his behalf. This key event occurs in an empty pub, the *Manx Fairy*, owned by Kate's acquisitive father, Caesar, on the evening of the signing of the fisherfolks' petition. Pete and Philip's roles as lover and husband seem to have been swapped round, and some aspects of their personalities interchanged. Heathcliff's guile and ruthlessness have translated into Philip's duplicity and all-consuming career ambition, while mild Edgar's incomprehension of Cathy's heart has found a substitute in generous Pete's complete disregard for Kate's feelings.

The lively Kate (played by the Polish-born actress Anny Ondra), who is the focal point of the whole film and the 'embodiment' of the *Manx Fairy*, prefers Philip, the restrained lawyer who is heir to a line of Deemsters,[171] to Pete, the broad-shouldered, broad-smiling and wide-eyed fishermen's leader (played by Danish actor, Carl Brisson). Pete (Heathcliff), who soon leaves for South Africa to seek his fortune and win the consent of Kate's father to their union, is mistakenly reported as 'killed'. Kate, freed from the promise she made to him, becomes involved with Philip but, predictably, Pete is alive and comes back a rich man – just like Heathcliff, in Chapter X, after the three years of his mysterious absence from Wuthering Heights.[172]

After having silently betrayed his friend, Philip also breaks faith with Kate whose attachment, if made public, could, in the opinion of his possessive and snobbish aunt, destroy the family's reputation yet again and ruin Philip's prospects of becoming a Deemster:

> Your father married beneath him – let his ruined career be a warning to you. [...] I've devoted my life to fitting you for the day when you become Deemster.

Philip is thoroughly aware of his aunt's class prejudice against the 'publican's daughter' and ashamed of his own conduct but fails to make amends – even when Kate tells him that she is expecting his baby, and gives birth to his daughter. Philip is thus more flawed than Brontë's Edgar who acquires depth after the simultaneous death of his wife, Cathy, and the birth of his daughter, Catherine, whose education he takes 'entirely on himself'.[173]

Despite the disgrace that the truth finally brings upon Philip and Kate, and the pain that the naïve Pete is made to feel, Kate will actually secure a future where her baby Katherine (cf Catherine) can enjoy her real father – and does not have to endure being motherless. Thus, Kate, in persevering with the more fallible of her two childhood friends – and recording, like Brontë's Cathy, the course of events in a little diary – could be seen as actively protecting herself from a potentially fatal entanglement with her wrong animus.

A major key to understanding the mythocritical nature of *The Manxman* resides in seeing the framing of the protagonists through the apertures of the film set, or the openings of the natural scenery, as a valid type of sophisticated montage. The protagonists resemble little figurines cut out of an animated backdrop. I am thinking of the 'scooped-out' rock that frames an exuberant Kate, up above in the sky, as she climbs up then swiftly comes down the cliff for her rendez-vous. The very same rock then frames her counterpart, a miniature Philip, down below, standing still against the animated backdrop of the sea, as he is anxiously waiting to break the news of Pete's return to her. Strikingly, the 'scooped-out' rock that frames Kate then Philip, will capture in the exact same manner the white sailing boats as they venture towards the wild cliffs of the Scottish island of Foula, in Michael Powell's *The Edge of the World* (1937).

This type of framing-editing, which would be stamped forever on Powell's memory and eyes after he still-photographed *The Manxman*, blends the aesthetics of the film with its thematics of entrapment. Christine Gledhill, in just two sentences, underlines the emotional impact on the public:

> The familiar if illogical plot twists generate emotionally piquant situations between the trio, orchestrated into telling pictures and scenes, framed by windows, doorways and arches. [...] Flicker-book montage between frontal close-ups sandwiches the spectator between demanding or heart-reading protagonists, whose glances meet ours as they direct their gaze to the other behind the camera and the viewing audience.[174]

Conversely, some inanimate components of the backdrop such as the scooped-out rock, the grinding stones of the old mill-house, the gaping fireplace at Kate and Pete's little fisherman's house, the flashing lighthouse or even the bright sunlight that illuminates the wood clearing of the lovers' innocent walks, because of their proximity to the characters, can fleetingly appear humanised and display empathy, hostility or both. The cinematography of *The Manxman* seems to be heralding four classics of the 'poetic-realist' era, Jean Vigo's *L'Atalante* (1934), Michael Powell's *The Edge of the World* (1937), Jacques Prévert's *Remorques* (1939) and Michelangelo Antonioni's *Gente del Po* (1943). In all four of those moving pictures, a naturalistic setting dominated by the presence of an untamed (or half-tamed) element (sea or river) is necessary for the drama to unfold, and for the characters to apprehend, in true Wordsworthian spirit, the 'primary laws of their nature'. The 'Natural Supernaturalism', in the words of Anne Williams, which permeates Emily Brontë's *Wuthering Heights*, could thus be seen as re-surfacing, in cinematic form, in the 1930s, on the incoming wave of 'Poetic Realism', which, supposedly, hit France first. *The Manxman*, conceived earlier on the other side of the Channel, with its humanised maritime landscape, its figurine-like characters and its sociopolitical tinges, may well have been a forerunner of the trend – and Albert V. Bramble's *Wuthering Heights*, the missing link between 'Natural Supernaturalism' and 'Poetic Realism'.

*   *   *

The aestheticism of *The Manxman* is characterised by a constant interplay between the changing emotions of the protagonists and the varied terrain of the Isle of Man, which runs the gamut of sceneries from the quiet leafiness of the inland woods to the ruggedness of the windy cliffs and seashore. The contrasting dwellings of Philip's family (The Christians), an isolated manor (compare Thrushcross Grange), and of Kate's parents (The Cregeens), the harbour's pub (compare Wuthering Heights), point to a class hurdle to be overcome and symbolise the lovers' difference of rapport with the outside world. The old mill-house where Philip and Kate meet secretly could itself be read as a translation of Heathcliff and Cathy's sanctuary at Penistone Crags, and would most probably inspire David Halliwell for the scripting of his erotically charged Wednesday Play, *Cock, Hen and Courting Pit* (1966) starring Maurice Roëves and Nicola Pagett.[175] Subsequently, Philip and Kate's mill-house will be 'desecrated' by the wedding celebration of the wrong couple, Pete and Kate.

The night tableaux involving their family homes – then Kate and Pete's cottage – amplify the protagonists' dilemmas while the pulsing beam of the lighthouse, falling incessantly on the narrow streets of the fishermen's village, adds a tragic and ominous quality to the *mise en scène*. This 'powerful moral chiaroscuro', in the words of Mary Hammond, originates in the Victorian tradition of serialisation and illustration of novels, and 'foreshadows the *noir* techniques of early European film-makers':

[...] despite the adherence to a strong Victorian pictorial tradition, The Manxman demonstrates that it is part of the new century: post-Great War, acutely conscious of the Depression and of more ambiguous moral values. The settings dwarf the actors. The framing is capable of trapping or excluding anybody. The pulsing lighthouse is both reminiscent of nineteenth-century tropes and at the same time disturbingly innovative, since it occurs relentlessly, almost without reference to the action, presaging nothing and yet everything.[176]

The prolific, copyright-watching author of *The Manxman*, Hall Caine, much appreciated by the pre-Raphaelite artist Gabriel Dante Rossetti – and by the successful Bram Stoker who dedicated his *Dracula* to him – had been a very famous novelist in his time. *The Manxman* alone had produced numerous stage plays until the film version of George Loane Tucker (1916), now lost, in effect took over from the theatre's adaptative trend. The new niche of cinema adaptation had a greater potential for a popularity-obsessed novelist whose writing years were over.

The adaptation of 1929 raises the question of adequate translation and shared authorship at many different levels since 'neither Hitchcock nor producer John Maxwell were happy with the film' although it was both a commercial and critical success, and was described in *The Bioscope* as a film of 'remarkable power and gripping interest'.[177] Is it Stannard – or both Stannard and Hitchcock – who decided to revert to the more subtle open ending imagined by Sir Hall Caine for his 1894 novel? The spectators, like Caine's readers, are simply left to imagine how Kate, Philip and their baby girl fare after their gruelling departure from the village. They are also left to imagine a future of seafaring adventures for the endearing Pete.

Here is the movie's last sequence, put into words by Christine Gledhill:

Cinema, seeking out and realising the incipient movement trapped in the picture, offers a low-angle long shot as Kate and Philip progress slowly uphill with the baby under a stormy sky. This dissolves into a big close-up of Pete's face at the mast of his boat, hair blowing in the breeze, his eyes softly grazing the camera lens: a mirror shot of the opening, except that now his boat is not coming home but setting out to sea.

The Manxman is at the heart of a natural process of story recycling, and constitutes a fine example of 'cultural intertextuality' since it is in the film by Stannard and Hitchcock, rather than in the novel by Caine, that the 'Natural Supernaturalism' of *Wuthering Heights* resides. Both Sir Hall Caine (the 'author') and the future Sir Alfred Hitchcock (the 'auteur') were keen to assume a 'trademark public persona' that simply allowed them to obtain full credit for some filmic creations born of a recasting of narrative, and existing only through a cooperative process. [178]

The team effort that Stannard defended in his speech to the *Kine-Cameramen's Society* in 1921, and most probably experienced with Bramble on *Mr. Gilfil's Love Story* and *Wuthering Heights*, echoes the work ethic of the continental film-makers of the poetic-realist wave, whose pictures were the fruit of a genuine cooperation between the screen or dialogue writers, the director, and the members of the crew and cast. Such was also the team spirit that the Anglo-Hungarian team, Michael Powell and Emeric Pressburger, known as 'The Archers', strove to incarnate some years later in the United Kingdom as they brought together the (superficially) irreconcilable trends of Poeticism (or melodrama) and Realism.

*   *   *

In the absence of footage, the archaeological exploration of the Bramble's *Wuthering Heights* has allowed an **incursion** into the work practices, partnerships and publicity material that both made and surrounded this silent film as well as *compensated*, to a certain degree, for the unfeasibility of a classic study of form and content: it has positioned Bramble's re-evaluated silent film into the incipient wave of *transnational* Poetic Realism. More 'mythocritically' maybe, this journey in time has rendered possible a re-invention, via *Mr. Gilfil's Love Story* and *The Manxman*, of the lost *Wuthering Heights'* cinematic themes, style and visual grammar, and foster a re-emergence of those dynamic structures of the Imaginary (thematical and aesthetical) that once engaged the **trust** of the production company, Ideal, and the strong interest of the Bramble-Stannard duet.

In this first British film adaptation of Emily Brontë's balladic novel, the phases of *appropriation* and *compensation* described by George Steiner in *After Babel* cannot be clearly identified due to the film's ephemerality and the rise, one generation later, across the Atlantic, of the highly influential cultural translation produced by Goldwyn (1938-1939); this Hollywood production would be immediately reclaimed and **assimilated** by the British critics and audiences. As our aim is to continue our *Journey Across Time and Cultures* chronologically and *transnationally*, we will need to direct our attention towards the Buñuel (not the Wyler) film first. This production takes us to the Paris and Madrid of the 1930s, and even to the Mexico of the early 1950s. We will then cross the terrestrial border to Goldwyn's California, from where we shall make a safe passage back to Europe, and linger within the confines of the United Kingdom. This return to the original soil gives rise to a cross-generational tour encompassing not only the 1939 classic – and contrasting British period dramas (1970, 1992, 1998 and 2009) – but also the television plays and *Classic Serials* and, in between, the unmade film envisaged by Lyndsay Anderson and Richard Harris (1963-1965).

As our perspective now shifts to the extant translated texts, the hermeneutics of **incursion** (or *appropriation*) and **compensation** will focus on the adaptation of the Gothic architecture of *Mise en Abîme* to its new milieu – and on the degree of involvement (or 'foreignisation') of the extra- and/or intra-diegetic observers/viewers. We will witness Heathcliff's sex change on the Yorkshire moors, and be momentarily 'stranded' on the un/familiar 'garrigues' of the South of France. We will then be ready to travel to the Far East and medieval Japan, and be dropped in the middle of a bloody 'Chambara'. A transformed Yorkshire landscape where traditional critical landmarks have faded away will re-open up straight away with Andrea Arnold (2011). This last discussion shall help in unveiling more completely still those persistent translative schemata and dynamic structures of the Imaginary at work in the subversive *Wuthering Heights* 'literary classic'.

Part III

The Heritage and Cross-Heritage Transformations: Aiming for the Ideal Re-Surfacing of the Dynamic Structures of the Novel

A

The Transformations of *Wuthering Heights* into the Contrasting Versions of Buñuel and Wyler

Chapter 21

L'Amour Fou as Hermeneutic of *'Incursion'* (1933–1953): A Collaborative Screenwriting and Shooting Practice (Buñuel-Unik-Jiménez) *'Re-Appropriated'* by Buñuel's *Mise en Scène*

Wuthering Heights (*Abismos de pasión*)
Directed by Luis Buñuel
Adapted by Luis Buñuel, Pierre Unik, Georges Sadoul and Jean Grémillon
  (1933 and 1936)
Re-adapted by Luis Buñuel, Arduino Maiuri and Julio Alejandro de Castro
  (1953)
Featuring Irasema Dilián and Jorge Mistral
90 minutes
Released in Mexico on 30th June, 1954
Released in France on 12th June, 1963
Released in New York, US, on 27th Dec, 1983

The wish expressed by Eliot Stannard to see the 'kind of trinity' embodied by the screenwriter, director and cameraman become a reality in the practice of modern film-making is not a wish that has been generally realised, unless the screenwriter, director and cameraman were one and the same person.[179] However, when broken down into the 'screenwriter-director' and 'director-cameraman' pairs, Stannard's collaborative dream can be said to have effectively materialised, one-half at a time, in a variety of cultural settings, which often had for common goal to facilitate a cross-heritage transformation. The contrasting versions of *Wuthering Heights* that succeeded Bramble and Stannard's silent version of 1920, at different times (1939 and 1953) and in different countries (Hollywood and Mexico), actualise this 'screenwriter-director-cameraman' trinity while helping to retrace the influence of some potent figures, such as the producer and the actors, on the final meanings of each film. They finally showcase the impact of the foreign soil on the dynamic structures of *Wuthering Heights*.

Stannard's trinity took many shapes during the uneven conceptual phase of *Abismos de pasión*, a project that spanned twenty years (1933-1953) and underwent several cultural adjustments – or '*incursions*', in the vocabulary of George Steiner. Unlike William Wyler – who first envisaged shooting *Wuthering Heights* after the script of Ben Hecht and Charles MacArthur had reached the stage of being purchased by the best buyers-adaptators[180] – Luis Buñuel's interest in making his version originated in the Surrealists' adherence to the concept of *l'amour fou* as well as in their profound admiration for Emily Brontë's novel:

It's a film I wanted to make at the time of *L'Âge d'or*. It's a key work for Surrealists. I think it was Georges Sadoul who translated it. They liked the side of the book that elevates *l'amour fou* above everything – and naturally as I was in the group I had the same ideas about love and found it a great novel. But I never found a backer, the film languished among my papers and was made in Hollywood eight or nine years later.[181]

After his collaborative ventures with Salvador Dalí on *Un Chien andalou* (1929) and *L'Âge d'or* (1930), towards the end of his fruitful stay in Paris, Buñuel initiated a participative screenwriting experiment with his talented poet friend, Pierre Unik:[182]

There were several subgroups within the [Surrealist] movement, which had formed according to certain curious affinities. Dalí's best friends were Crevel and Éluard, while I felt closest to Aragon, Georges Sadoul, Ernst, and Pierre Unik. Although Unik seems to have been forgotten today, I found him a marvellous young man, brilliant and fiery. [...] Pierre published two collections of poetry and edited the Communist party journal for children.[183]

Another who came to join in the Paris experiment was the film critic and historian, Georges Sadoul:[184]

It was in this year [1933] that he first decided to film *Wuthering Heights*. His collaborator on *Las Hurdes* [1932], Pierre Unik, [...] helped with the adaptation. Georges Sadoul also worked for a night on the script. The choice of *Wuthering Heights* was significant. André Breton and the Surrealist group had adopted the book as one of their banners, along with the liking for the early English 'Gothick' novels. The review *Minotaure* had dedicated many pages to it, with illustrations by the best Surrealist painters of the period [Balthus amongst them].[185]

When Buñuel became an executive producer for *Filmófono*, a production company aiming to produce some commercial feature films in a studio environment as strong and profitable as the Hollywood model, and whose commercial strategy relied heavily on the adaptation of high-profile Spanish literary works, the experiment on *Wuthering Heights* could continue on the side while taking on some Spanish overtones. In Francisco Aranda's critical biography still, the French poetic-realist film-maker, Jean Grémillon, an enthusiastic hispanophile proficient in Spanish, reminisces how he and Buñuel started reworking what, depending on the decade, Buñuel either called 'a first-rate script' – meaning a 'screenplay'[186] – or 'a twenty-page long synopsis' – 'une trame narrative d'une vingtaine de pages'.[187] The fresh ideas that Grémillon grafted on to the French Surrealist text led to the first translation of the initial script (*Les Hauts de Hurlevent*) into the Spanish cultural context (*Abismos*[188] *de pasión/Depths of Passion*, and to its complementary title, *Cumbres Borrascosas/Wuthering Heights*). Grémillon, however, was forced to flee Madrid in the Summer of 1936, and his name ceased to be associated with the project. Buñuel also returned to Paris but soon embarked for Hollywood, then New York.[189] The two friends pursued separately their careers of politically aware film-makers, also able to set a particular cinematic trend and write movies appealing to a wide audience. In France, the avant-gardist Grémillon soon became recognised as a popular film-maker by making a series of three commercial hits characterised by a certain 'realist lyricism', *Gueule d'amour* (1937) with Jean Gabin, *L'Étrange Monsieur Victor* (1938) with Raimu and *Remorques* (1939-1941) with Jean Gabin and Michèle Morgan. As to *Wuthering Heights*, it was put on hold – and Buñuel never expanded on the first adjustments that had rendered Pierre Unik's synopsis suitable to the Mexican context.

In order to materialise on to a cinema screen, *Les Hauts de Hurlevent/Abismos de pasión* had to wait for Buñuel to be established as a film-maker in Mexico, and awarded the Best Director prize for *Los Olvidados* at the 1951 *Cannes Film Festival*. His producer, Óscar Dancigers, would back the adaptation of *Wuthering Heights* on the condition that Buñuel agreed to use a pre-existing cast, Irasema Dilián[190] (Catalina/Cathy), Jorge Mistral (Alejandro/Heathcliff), Lilia Prado (Isabel/Isabella), Ernesto Alonso (Eduardo/Edgar) and Luis Aceves Castañeda (Ricardo/Hindley):

> [...] Pierre Unik and I had written a screenplay based on *Wuthering Heights*. Like all the Surrealists, I was deeply moved by this novel, and I had always wanted to try the movie. The opportunity finally came, in Mexico in 1953. I knew I had a first-rate script, but unfortunately I had to work with actors Oscar had hired for a musical – Jorge Mistral, Ernesto Alonso, a singer and rumba singer named Lilia Prado, and a Polish actress named Irasema Dilián, who despite her Slavic features was cast as the sister of a Mexican métis [Luis Aceves Castañeda]. As expected, there were horrendous problems during the shoot, and suffice to say that the results were problematical at best.[191]

Julio Alejandro de Castro, a brilliant Aragonese poet in exile in Mexico, added to the list of high-profile contributors to the script (Unik, Sédoul and Grémillon) and initiated a long-lasting collaboration with Buñuel, since he later co-wrote with him *Nazarín* (1959), *Viridiana* (1961), *Simón del desierto* (1965) and *Tristana* (1970). The Italian screenwriter, Arduino Maiuri (Irasema Dilián's husband) also appeared in the writing credits and would have helped flesh out a script that needed to be ready quickly as well as be detailed enough for the actors to face, with relative confidence, a tight shooting period of twenty-five days.[192]

In France, as the poetic-realist movement was emerging from the thought-provoking Surrealist era, Buñuel had been denied the opportunity to shoot *Les Hauts de Hurlevent* and complement thematically his avant-gardist first feature, *L'Âge d'or*, with the perfect literary modernisation, both reputable and fashionable. Who did he then have in mind for his cast? This will probably remain a mystery. The film project of *Les Hauts de Hurlevent* was not as advanced as Spanish film critic and Surrealist author Francisco Aranda has suggested in his remarkable biography on Buñuel.[193] Further, Luis Buñuel never gave a hint about the actors he would cast or, at least, refrained from doing so before *Abismos de pasión* had grabbed the attention of the New Wave critics. This only occurred more than a decade after the film had been made!

In the late 1960s, his comments on the subject were prompted by a question from the Mexican film critic, Tomás Pérez Turrent, which rounded off a rather difficult interview. As plainly revealed by Saviour Catania in 'Wagnerizing *Wuthering Heights*' (2008), Buñuel could keep tight-lipped, or even act playfully inconsistent with his interlocutors. The mischievous director thus conceded to Pérez Turrent that he could see himself re-shooting *Les Hauts de Hurlevent* in England (or in France) 'with a good cast' including Claudia Cardinale (as Catherine);[194] he had been impressed by her performance in Luchino Visconti's *Sandra* (1965).[195]

Claudia Cardinale, like Irasema Dilián before her, would have impersonated a Catherine/Catalina nearing thirty, and Buñuel's re-vision of his own film – by sticking to the age of the lovers in *Abismos de pasión* – would have, again, transformed the unconscious of Emily Brontë's novel. In the novel, Catherine is fifteen years old when Heathcliff, aged sixteen, abruptly leaves Wuthering Heights, then eighteen when he suddenly reappears, three years later, at Thrushcross Grange, 'on a mellow evening in September'.[196] At the time of Heathcliff's return, she would not have been married to Edgar Linton for more than five months and, in March of the following year (which would be the year 1784), she dies while giving birth to her daughter, the second-generation Catherine – not to the baby son of the Buñuel version, who vanishes at the same time as his mother Catalina.

Cathy/Catherine is thus this young girl of eighteen who welcomes back her lost childhood companion, Heathcliff, and dies a mere six months later – the tragic interval that Buñuel and his co-scenarists exploited for their highly condensed treatment of the story. How does this ellipsis – which does much more than remove the second generation of lovers – and the change in the ages of Cathy and Heathcliff (and, consequently, in the ages of Maria/Ellen and Cathy's brother, Ricardo/Hindley) affect the Mythical Components (MCs) and Bataillan Themes (BTs)?

At this point, it is useful to introduce another key concept of George Steiner, that of **'betrayal by augment'**:

> Certain texts or genres have been exhausted by translation. Far more interestingly, others have been negated by transfiguration, by an act of *appropriative* penetration and transfer in excess of the original, more ordered, more aesthetically pleasing. There are originals we no longer turn to because the translation is of a higher magnitude (the sonnets of Louise Labé after Rilke's *Umdichtung*). I will come back to this paradox of **betrayal by augment**.[197] [Emphasis added]

This concept will be particularly useful in relation to both the Buñuel and Wyler versions.

*   *   *

As a result of the considerable shortening of the narrative, the Gothic Figure of Lineage snaps, and the associated architecture of *Mise en Abîme* is flattened. There is no room for the mediation of a Lockwood and, with the intra-diegetic narrators being removed, the extra-diegetic observers (or film spectators) have little opportunity to get close to the protagonists. Even the eroticism is distant and self-contained, present at character level only in Catalina, Isabel and Alejandro. Notwithstanding, the *Mise en Abîme* three-dimensional Figure is recovered matter-of-factly, in the introductory sequence to the movie through the motifs of the bird cage, and of the butterfly jar and vitrines. Further, it is recovered symbolically, with a denser emotional charge, in mid-film, through the metaphoric image of the sheltering canopy of Catalina and Alejandro's childhood tree, and finally, at the end, through the striking image of the steep flight of stairs descending into the crypt, where Catalina's body lies in its coffin.

The outline of Buñuel's Mexican *Wuthering Heights* is flat and compact; that major structural idiosyncrasy – or departure from the novel – comes with its own trail of aesthetical and thematic surprises. In a necessary inversion, though, the world of La Hacienda/Thrushcross Grange borrows a crucial component from La Granja/The Heights, the farm animals, which imply, in the context of this arid and dusty land, a big estate mentality, and a certain ruthlessness. The pig, in particular, seems to be slaughtered whimsically rather than for the meat it brings, which reinforces the theme of Eduardo's 'civilised' violence and Isabel's untenable callowness. Catalina's marriage into this lame family seems to meet the simple purpose of injecting new blood, and bearing Eduardo a son.

In order to assess the cross-heritage transformation of the MCs and BTs, the introductory sequence, which announces the final one, is worth commenting upon. It is truly 'sensory', as auditory as it is visual in its precise unfolding. The firing of a gun triggers the flight of the vultures perched on the leafless, skeleton-like branches of a dry tree. Isabel, frightened by the tumult, runs back into the hacienda where her brother Eduardo is meticulously pinning down a butterfly for his collection. She reveals the identity of the 'rifleman' who has been concealed off-screen: Catalina. The latter soon makes her entrance with her own rifle; she killed one of the vultures with a single shot.

In this opening sequence, the trio Catalina, Isabel and Eduardo is studied through their different rapports with the wild (the birds of prey) and the tame (the caged bird and the dead butterflies) as well as with the strong (those wielding a firearm) and the weak (those cringing at rifle shots or inflicting a slow, painful death on small creatures). Those rapports of strength condition the way they objectify and see love, hence the disenchanted tone of *Abismos de pasión* – and of the related closing sequence in the crypt.

On the very morning of his comeback at La Granja and fight with Isabel, who is now his wife, after a night spent expecting the news of Catalina's death, outside La Hacienda, Alejandro hurries to the graveyard where Catalina's body has been put to rest in the family crypt. He is followed by Ricardo, Catalina's brother, who carries a rifle. Alejandro forces the trap doors to the crypt open, and is shot in the heart by an off-screen rifleman. Despite his wound, he takes the narrow staircase leading down to the crypt and opens up Catalina's coffin: she is dressed as a bride and Richard Wagner's strings are more audible than ever. After he kisses her, he hears her calling his name, behind him, on the stairs. He sees her, she smiles, and her arms are wide open. A split second later, Alejandro seems to fleetingly realise that Ricardo is aiming a rifle at him, from the narrow staircase. Then, in a (cinematographic) reverse shot, Alejandro falls dead on Catalina's body. In yet another reverse shot, Ricardo climbs out of the tomb and closes down the trap doors; the end credits already appear. At this stage, it is not clear who, amongst the extra-diegetic observers, has managed to escape with Ricardo from the lovers' tomb (*Abismos de pasión*) or who, amongst the film spectators, is willing to follow Ricardo in the disenchanted world dreamt by Buñuel.

In this last sequence, the survival instincts based on self-interest and incarnated by Ricardo – the Evil Morality – dominate, and there is no coming back from the tomb if Eros allured you there. The Bataillan Interdicts related to Hyper-Morality and Sexuality (the Second and the Third, respectively) are profoundly altered, for there is a kind of fateful inertia at intra-diegetic level, and no possibility of a rewarding trespassing experience that would lead to learning. As a consequence, the Third MC (MC3) connecting Love to the Sacred is put into question, then hidden away in an unforgettable crypt.

This vision of the illuminated crypt is overwhelming, being profoundly oneiric. The surrealistic script and impeccable cinematography 'capture' the stark motif of the grave – which is embedded in Emily Brontë's *Wuthering Heights* as well as in her poem, *Remembrance*[198] – and 'betray by augment' the Gothic images present in both:

> By placing Caterina[199] in an underground crypt, Buñuel's final scene captures Brontë's image of having the church thrown down over Catherine.[200] […]. In essence Caterina is not a bride of life, but of death. As in the novel, she has haunted Alejandro who literally dies with her underground. For Brontë the 'mad love' uniting Catherine and Heathcliff can be resolved only in death, just as for Buñuel and the Surrealists' passionate love is not possible in this world; it is a form of death, an annihilation of everything beyond the confines of the lovers.[201]

The mythical journey is truncated. There is no room for joyful expiation[202] – and the single-generation structure prevails despite the mention of Catalina/Catherine's baby son, and the appearance of her little nephew, Jorgito/Hareton, in the sordid world of La Granja. Passion is sterile, in the cinematic confines of *Abismos de pasión*, inasmuch as both children are transparent creatures who, at the time of the lovers' death, disappear totally from the screen and dialogues, seemingly taken away by Thanatos herself:

> In the novel by Emily Brontë, love has its basis in a mystical communion of identical souls who bring on disaster if they abandon one another for worldly trifles – but a disaster which endures only until the next generation. In the film by Luis Buñuel, love is based as much on aggression as on communion, and disaster endures forever.[203]

It follows that the Second Mythical Component connecting the Living and the Dead, and associating Childhood and the Sacred, is profoundly modified after the foreign *incursion* made by Buñuel into the unconscious of the novel. Further, from José/Joseph's myopic point of view – and, maybe, from Buñuel's clear perspective as *metteur en scène* – there is no possibility of fulfilment for love in the afterlife:

> And our name in time shall be forgotten, and no man shall have any remembrance of our works.
> For our time is as the passing of a shadow, and there is no going back of our end: for it is fast sealed, and no man returneth.
> Come therefore, and let us enjoy the good things that are present, and let us speedily use the creatures as in youth.[204]

* * *

The Present thus described is an aggressive Carpe Diem, which is motivated by material gain and which makes no sense of the Sacred. Conversely, youth – and Childhood – is dehumanised and cut away from adulthood, both devalued and hunted for as the highest prize. In *Abismos de pasión*, the First BT associating 'Childhood, the Present and the Sacred' is thus next to being unrecognisable. Moreover, the Present does not constitute either a stepping-stone for re-generation through Discovery, or a stepping-stone for self-revelation through Imagination. In the liminal space of the tomb, Alejandro – the closest character to being an intra-diegetic outsider – hallucinates the present moment and is deluded by the phantom of his past love. Still, he is the one who Trespasses and introduces to the film spectators the new and hostile territories of La Granja/Wuthering Heights, and of the Crypt.

The Third Mythical Component (MC3), the Initiatory Love Journey, which is so much dependent on the *Mise en Abîme* architecture of the novel, is displaced within its own archetypal layer, and has transmuted into a Fatal Love Delusion, which, according to Julie Jones' article, 'Fatal Attraction' (1997), takes on an 'Oedipal' rather than a Romantic tinge:

> In his underscoring of the instinctual rather than the conventionally romantic, Buñuel seizes on a biological element that gets no play at all in *Wuthering Heights*: Cathy's

pregnancy. [...] The nesting impulse may explain why she [Catalina] is content simply to have Alejandro around in a non-sexual way. Oddly enough, the film goes even farther, suggesting a subtle shift in their relationship, which takes on a peculiarly *Oedipal tinge*. This transformation has no place in Brontë's scheme of things. If anything, the situation in the novel suggests incest between siblings. Although this idea enters into the film – early on Cathy describes Alejandro as 'más que un hermano' – Buñuel apparently finds the *Oedipal motif* more interesting.[205] [Emphasis added]

Throughout the film narrative, the fatalism of the Third Mythical Component is attuned to the pessimism of the metaphors carrying the First MC (MC1). For the connection between the Profane and the Sacred, through Mind and Nature, is well translated in *Abismos de pasión*. As noted by Anthony Fragola in his previously cited essay, 'Buñuel's Re-vision of *Wuthering Heights*: The Triumph of *L'amour fou* over Hollywood Romanticism' (1994), the atmospheric settings and images carry the theme of *l'amour fou*. The storms are consistently associated with Alejandro's 'inner torment', and the soft clouds materialise this privileged connection between the lovers:

> Buñuel's subtitle, *Cumbres borrascosas*, a literal translation of *Wuthering Heights*, conveys tempestuous storms at the heights of passion; coupled with the title *Abismos de pasión* (*Depths of Passion*) it represents the polarities associated with *l'amour fou*. For example, as Caterina [Catalina] wills herself to die, Alejandro stands vigil in the driving rain. Although melodramatic, Buñuel's use of storms reveal the characters' inner torment. [...]

> Clouds, another naturalistic detail in *Abismos de pasión*, serve as a self-referent to Buñuel's first two masterpieces *Andalusian Dog* (1929) and *L'âge d'or* (1930) and also echo Brontë's aforementioned naturalistic images of 'moonbeam', 'lightning', 'frost', and 'fire'. [...] José's [Joseph's] recitation from the *Book of Wisdom* wherein he uses naturalistic images, including 'and our life shall pass away as the trace of a cloud', further reinforces Buñuel's use of clouds to illustrate the connection between the lovers and the natural world.[206]

The topographical opposition between the Heights and Thrushcross Grange is translated into the contrasting layouts of the farmhouse ('La Granja') – which is dark and labyrinthine – and of the 'Hacienda' – which is light and spacious. Furthermore, the botanical symbolism derived from the Yorkshire setting has been transferred, for the entirety of the film, into animalistic and entomological metaphors that reveal the characters' psychology as well as highlight the inherited traits, religious superstitions and predetermined behaviours. Thus, Buñuel's animalistic and entomological metaphors overshadow with gloom the dream of a 'free and sovereign love' that was made by Emily Brontë and seen by Georges Bataille. These metaphors are usually conveyed through some documentary-like inserts, some sadistic vignettes that are mentioned by Pérez Turrent in his conversations with Buñuel,[207] and by Michael Popkin in his already cited '*Wuthering Heights* and Its Spirit' (1987) that illuminated, in Part I, the unconscious of the *Wuthering Heights* texts with its reference to *Beauty and the Beast*:

[...] as I understand the popular ballad and folk tale origins of the novel, they are replaced by Buñuel with his own vision of a love whose basis is cruel treatment of the beloved and sadistic mistreatment of everyone else. The element in the novel to which Buñuel is most faithful is the presence in it of animals who are tortured and gratuitously killed, an element which not only pervades the film but serves as a metaphor for the way the characters interact. [...] The world is partly that of the Brontë novel, in which Heathcliff hangs Isabella's dog, and where his counterpart in the next generation [Hareton] is offhandedly said to be hanging a litter of puppies. The viewer of Buñuel's film, however, is far less likely to think of Brontë than of other films by Buñuel, particularly *Viridiana*, in which a cat leaps upon a mouse as the heroine is subjected to even more indignities than Isabel (Isabella) in this version of *Wuthering Heights*.[208]

The unemotional close-up of the butterfly dying under Eduardo's magnifying glass typifies the detached, near-objective shooting style required from Agustín Jiménez, Buñuel's cinematographer. Like Gregg Toland for the 1938-1939 version, he would participate in creating a specific mood. His close-up on the dying butterfly enhances the ephemeral beauty of the insect while asserting the transience of human passions, and announces the sadistic touch of the following inserts. In the vibrant Mexico City of the 1930s, Agustín Jiménez belonged to the avant-garde of photographers. At that time, the likes of Henri Cartier-Bresson, Sergei Eisenstein and his celebrated cameraman, Edouard Tissé, lived in the city.[209] Prior to the opening, in 1931, of Jiménez's first one-man show at *The Galería Moderna*, Eisenstein firmly asserted, 'I believe that Agustín Jiménez is a great photographer and I appreciate his excellent work.'[210]

The presence on *Abismos de pasión* of an expert cinematographer who had also been photographing Buñuel's *El Bruto* (1952) served to alleviate the sense of discrepancy in the actors' physiques (and accents) when the film was first released in Mexico. In the long run, this casting mismatch appears to have introduced an element of unreality and atemporality that adds to the oneirism of the picture.[211] Further, the discrepancy in the actors' accents can easily be blocked out by the film spectators who live in a different time and generational zone, and even by a first-class film scholar like Kamilla Elliott, who would sometimes rely too heavily on the English subtitling and on the textual evidence found in her own argumentation:

In the final scenes of Buñuel's *Abismos de pasión*, Catalina's (Cathy's) physical resemblance to her brother, Roberto [Ricardo] (Hindley), is emphasized over any resemblance to Alejandro (Heathcliff), not only in terms of biological genetic structure, but also of narrative structure.[212]

The success of the Buñuel-Jiménez partnership is most tangible in the composition of those sequences that are built on some striking naturalistic metaphors (and archetypal images) situated in the dynamic structures of the novel. The archetypal images of the crypt and the tree used in *Abismos de pasión* reach into the deepest layers of the Jungian unconscious.[213] The cinematographer's vision of the primeval tree, in particular, with its reassuring canopy and gigantic roots, encompasses Emily Brontë's idiosyncratic visions of Penistone Crags – Cathy and Heathcliff's refuge on the moor – and of the 'rough sand-pillar' at the foot of which Nellie and Hindley have buried their childhood relics.[214] As noted by Anthony Fragola, this all-encompassing vision of the primeval tree connects the Profane and the Sacred (MC1), injects some cinematic life into the otherwise partially discarded Second MC (MC2; Correspondence between Childhood and the Sacred; Connection between the Living and the Dead), and 'serves as an archetypal image':

> Another naturalistic detail in *Abismos de pasión*, a massive tree, has no parallel in the Hollywood version. The novel is 'rustic all through. It is moorish, and wild, and knotty as the root of heath' [from Charlotte Brontë's 'Preface to the 1850 edition of *Wuthering Heights*']. To capture the essential image of the root Buñuel has Caterina [Catalina] and Alejandro return to the tree where they played as children. […] The tree, huge and ancient, spreading its roots like tentacles and burrowing deep into the earth, […] serves as an archetypal image.[215]

The end result of this translative *incursion* into *Wuthering Heights* is the unveiling of the mechanisms of a *betrayal by augment*, the intrinsic value of which is demonstrated by the continued popularity of *Abismos de pasión* with the cinéastes and film reviewers worldwide – Jacques Rivette among them.[216] The disenchanted tonality of *Abismos de pasión* would pervade Buñuel's subsequent studies in passion adapted from novels: I am thinking of *Viridiana* (1961), *Tristana* (1970) and *That Obscure Object of Desire* (1977).[217] Further, most of the cross-heritage transformations orchestrated by Buñuel correspond to some high-profile adaptations of a wide array of literary texts (Spanish, English or French) that are not only given an omniscient, often pessimistic, narratorial bent but also studded with some oneiric motifs rooted in the collective unconscious. To meet the challenge of cultural translation, Buñuel worked with some remarkable co-writers (such as Pierre Unik, Julio Alejandro de Castro and Jean-Claude Carrière) while leaving both scriptwriting and cinematography open to some additional creative influences – Georges Sadoul, Jean Grémillon and Agustín Jiménez[218] on *Abismos de pasión*. Luis Buñuel's cross-heritage transformations, inclusive of *Wuthering Heights*' dynamic archetypal transformation, could thus be typified as *films d'auteur(s)* and actualise vividly, in all contexts, Eliot Stannard's dream of a cinematic trinity.

Chapter 22

L'Amour Mercenaire as Hermeneutic of *'Incursion'* (1938-1939): A Collaborative Shooting and Editing Practice (Wyler-Toland-Mandell) *'Pre-Appropriated'* by Goldwyn and His Stars

Wuthering Heights
Directed by William Wyler
Adapted by Ben Hecht and Charles MacArthur (1939, Goldwyn)
Featuring Merle Oberon and Laurence Olivier
102 minutes; 9,180 feet
Premiered on 13th Apr, 1939, Hollywood and New York, US
Trade Show on 28th Apr, 1939, London, UK

With the 1939 *Wuthering Heights*, Eliot Stannard's dream of shared auteurship appears to be considerably abridged. The Hollywood-made movie re-shuffles and re-invents the trinity 'director-screenwriter-cameraman' into the triad 'producer-director-cameraman' by dissociating the work of the original literary adaptators (Ben Hecht and Charles MacArthur) from the work of the cinematic team. The initial phase of cultural translation does not belong any longer to the film-makers, who should be called (privileged) 'interpreters' rather than (pure) 'translators'. Nonetheless, the film-makers do run the show. Together with a high-profile script doctor and a dedicated film editor, they, not the authors of the original screenplay, see their pre-visualisation of *Wuthering Heights*' filmic tableaux materialise on film. Another kind of participative practice has taken shape and become a reality. This complex reality, which we are about to unfold, originates as much in the professional make-up of the director, William Wyler, as in the workings of the film industry, in Hollywood, at the time.

While he was never drawn into writing himself, Wyler, in his absolute certainty about the worth of Ben Hecht and Charles MacArthur's screenplay, imposed it on his producer Samuel Goldwyn, who could not ignore that he had an extraordinary flair on four consecutive feature films *Dodsworth* (1936), *Come and Get It* (1936), *These Three* (1936) and *Dead End* (1937), which are now Hollywood classics. Wyler, strong with his four directorial successes, left Goldwyn no choice but to acquiesce to his overwhelming wish of shooting Hecht and MacArthur's script:

> Wyler did have a script he wanted to direct: *Wuthering Heights*, based on Emily Brontë's 1847 novel, a Gothic tale about the doomed passion between Catherine Earnshaw and Heathcliff.

[…] 'It was a marvellous script. I took it to Goldwyn. He read it and didn't want to do it. He didn't like stories with people dying at the end. He considered it a tragedy. I told him it was a great love story.'[219]

In persuading a reluctant Goldwyn to 'buy the rights', Wyler was not only showing his strong appreciation for a well-crafted piece of cultural translation but also revealing his awareness of the dynamic structures of Emily Brontë's novel. He manifested a sympathy for the controversial themes, which the Hecht-MacArthur script could neither completely mask nor develop. In common with Luis Buñuel, he used to bring a lot of acumen into the initial screenplays and would breathe his own dissenting personality into them. While Buñuel claimed, 'I am only interested in human relations',[220] Wyler, who shared a profound liking for psychological subjects, declared:

I have never been as interested in the externals of presenting a scene […] as I have been in the inner workings of the people the scene is about.[221]

Despite the pressure of the studios (Universal and Goldwyn) that promoted the specialisation of tasks, and maintained a safe distance between screenwriting and directing, Wyler had developed early in his film career the habit of taking back control. He took part in the selection of the scripts and tweaked them at pre-production level. Throughout their compilation of critical essays entitled 'Dossier William Wyler' for *Positif* (June 2001), Michel Ciment and Christian Viviani put the stress on the dynamism of this participative practice.[222]

Wyler's partner to pep up the screenplay of *Wuthering Heights*, which began shooting on 5th Dec, 1938 (Chatworth, California),[223] was his high-profile friend John Huston who had already helped him in strengthening and personalising the screenplays of *A House Divided* (1931) and *Jezebel* (1938). As Goldwyn was faced with the staunch refusal of Huston, his scenarist, and Wyler, his director, to make Cathy and Heathcliff 'more likable', there would be, perpetually, an atmosphere of fight during *Wuthering Heights'* pre-production conferences.[224]

Wyler, in a much more appeased state of mind, used to discuss the emerging screenplay at length with his cinematographer Gregg Toland, just as he had discussed with him their earlier films back in 1936-1937, and would discuss again the rest of their hugely popular collaborative work, *The Westerner* (1940), *The Little Foxes* (1941) and *The Best Years of Our Lives* (1946).[225] For each picture, Wyler and Toland agreed on a joint shooting strategy:

The pair soon developed one of the most fruitful director-cinematographer relationships in film-making history.
    […] 'When photographing a thing, he [Gregg Toland] wanted to catch the mood,' Wyler explained. 'He and I would discuss a picture from beginning to end. The style of photography would vary, just like the style of direction. […] We would discuss the style

of photography that would fit the picture, then the style of sequence. I would rehearse and show him a scene. Then we would decide together how to photograph it.'[226]

For *Wuthering Heights*, which Toland defines as 'a love story, a story of escape and fantasy', the pair was aiming at a subtly dreamy and unreal atmosphere, 'a soft picture, diffused with soft-candle-lighting effects'. Toland would emphasise the 'beautifully' romantic in his close-ups on Cathy/Merle Oberon's and Heathcliff/Laurence Olivier's faces, which 'could be kept in partial darkness, then come into the light at climactic moments'. For the indoor scenes at the Heights, he also advised Wyler to keep the cameras in a low position, so as to 'capture the ceilings of the sets' and accentuate 'the stifling confines and dour loneliness' of the place.[227]

Further, James Basevi, the Art Director, who would be nominated at the 1939 Academy Awards for his work on the film,[228] had re-imagined the contrasting pair of houses by designing their interiors as well as building their external frames in the Goldwyn studios. Basavi had also re-created the Yorkshire moors, on location, in Chatworth. The grandiose set, together with the subversive force of Toland's camerawork and Wyler's *mise en scène*, resulted in the memorable re-creation of, in Leonard Maltin's words, an atmospheric 'chiaroscuro country of the mind'. The level of presence of the First MC (MC1) was very high:

> The setting for the film was not the moors of Yorkshire, but a wilderness of the imagination. To have reproduced on the screen any large expanse of landscape would have been to chain the story and its characters to the actual. Instead, Toland and Wyler devised a close-in camerawork which, in every shot, seemed to show only a small part of the whole scene, in which roads, crags, housetops, and human figures were revealed in outlines against dense grays and blacks. Thus was created a *chiaroscuro country of the mind* in which the passionate Brontë figures can come credibly alive.[229] [Emphasis added]

*   *   *

The connivance between Wyler and Toland would also facilitate the implementation of the more consensual changes that had been decided upon during the difficult conferences with Goldwyn. In Barthes' narratological terminology, those changes to the original screenplay related to the Distributional Functions,[230] in particular, the catalysing of the 'Cardinal Actions and Events' or the catalysing of the 'Plot':

> The strain of sacrificing pictorial comment to exigencies of plot may be further illustrated by the number of discrepancies between the shooting-script and the movie's final print. Not only are many of Emily Brontë's cinematic details abandoned but also many that were originally proposed by the screen writers.[231]

More significantly still, their complicity would prolong into the shoot itself Wyler's counter tactics to Goldwyn's mainstream approach to the novel's dynamic structures or, in Barthes' narratological terms, to the Integrational Indices: representations of place, notations of atmosphere and psychology of the characters. In *Wuthering Heights*, those Indices are assuming the role of Functions inasmuch as they are intertwined with the three-dimensional Figure of *Mise en Abîme*, which is unaccounted for in the narratological theories of film adaptation, but preponderant in the Gothic genre.

Wyler and Toland's aesthetical tactics were supplementing the movie that Goldwyn had in mind with the energy of the unconscious of Emily Brontë's novel. In this respect, Goldwyn, in tune with the fashionable commentators of the time, incarnated a necessary, although particularly challenging obstacle. As highlighted by Pamela Mills in her essay, 'Wyler's Version of Brontë's Storms in *Wuthering Heights*'(1996), he held some consistent beliefs about love stories, their happy endings and their attractive, right-minded heroes:

> William Wyler seems to have been hampered by Samuel Goldwyn, the producer who not only wanted to sell films but who also had preconceived ideas as to how conventional heroes and heroines were to behave.[232]

During the shooting of the movie, the director had to rely on an additional ally, the film editor Daniel Mandell, who had been associated with the Wyler-Toland team on *These Three* (1936) and *Dead End* (1937), then would brilliantly collaborate with them again on *The Little Foxes* (1941) and *The Best Years of Our Lives* (1946), the latter film winning him his second Academy Award. In his remarkable biography on Goldwyn (1989), A. Scott Berg retells how Danny Mandell 'saved the day' when an irate Goldwyn came to question the time and money spent on finding the right shooting style for *Wuthering Heights*:

> After three weeks of shooting, [...] He [Goldwyn] found Wyler guilty of overshooting and overdirecting [...] but Wyler explained that he was trying to create a mood in *Wuthering Heights*, which could come only from shooting unusually. He repeatedly told Goldwyn 'not too worry, that it would all piece together'. [...]

> Wyler got Mandell to work overtime so that Goldwyn would view as few unedited sequences as possible. The producer thought it 'disgraceful', for example, when he saw the many camera angles from which the simple scene of Heathcliff's throwing himself on his pallet in the stable and thrusting his fists through the windowpanes was photographed. [...] [But] when Goldwyn saw the best of those shots spliced together, drawing in on an extreme close-up of Heathcliff, he appreciated the intense emotional effect Wyler had created.[233]

Wyler had been seeking and creating new cinematic solutions for the characters' immoral repressed feelings, in particular, Cathy's, to touch the audience, and for the meaningful variations in atmosphere and mood to show on screen. His aim was to allow the dynamic structures of the Imaginary, reconstructed patiently with the right cinematographic combination of Art Design and Integrational Indices, to contain the strong currents of *mercenary* love coming from the Hollywood template of 'the stable boy and the lady', and saturating the dialogues of *Wuthering Heights*. Goldwyn eventually came to see the positive effect that this type of *mise en scène* had on the actors' performances, and appreciated fully the emotional hold it had on the viewers of the carefully edited 'rushes'.

This prodigious producer used to anticipate, as well as gauge fairly well, the expectations of the North American audience living on the West Coast. Thanks to some well-organised test screenings, he could monitor precisely its specific reactions to a given picture. Starting with his initial reading of Hecht and MacArthur's screenplay, he had sensed the moral pitfall of *l'amour fou* in the context of a Puritanical society, and was aware of the commercial pitfall of its unedulcorated translation on to the cinema screens. To be perfectly safe, clear and comprehensible, the screen translation of *Wuthering Heights* required a *Mise en Abîme* (Ellen/Flora Robson, the intra-diegetic narrator) and the rather theatrical apparition, at the end of the picture, of the 'ghostly' figures of Cathy and Heathcliff. These very last additions to the script seem to obey the logic of the Second and Third MCs (MC2 and MC3):

> Just two weeks after the last retakes had been shot, *Wuthering Heights* sneak-previewed in Riverside [...]. The audience's questionnaire cards were among the worst responses to a motion picture he [Goldwyn] had ever read. They found the story hard to follow and seemed to concur with Goldwyn's initial instincts about to the material. [...] It occurred to Jock Lawrence [the top executive in charge of Goldwyn's publicity] that Flora Robson, whose character was relating the story of Cathy and Heathcliff to a traveler [Lockwood], was still in town. He thought she could read several short, lyrical speeches that might be dropped in at the half-dozen confusing junctures. [...]
>
> If *Wuthering Heights* was such a great love story, Goldwyn saw no reason why Cathy and Heathcliff could not be shown as ghostly figures, united at last, walking hand in hand to heaven.
>
> [...] He took the doctored film to Santa Barbara for a second preview, and a burst of applause drowned out the last bars of the music over the end credits. 'Well', Goldwyn remarked to the team huddled around him, 'they understood it.'[234]

In common with *Abismos de pasión*, *Wuthering Heights* possesses this admirable and ambiguous quality of being a cultural translation that outweighs Emily Brontë's text. If Goldwyn's last remedial choices inadvertedly re-connected with the dynamic structures of the novel, they were nonetheless also undermining Wyler's strategy and, more precisely, the dark restraint he had used in his *mise en scène* to convey the silent plight of Cathy/Oberon, torn between passion and restraint, Hyper-Morality and Morality.

For the vast majority of the Anglo-American film critics, from George Bluestone to Charles Higham and Pamela Mills, the Wyler picture is believed to supersede the novel in intensity and popularity. However, those commentators who have looked up to the Buñuel movie, from Jacques Rivette to Anthony Fragola and Julie Jones, have often felt compelled to round off their own discussions on adaptation by looking down on the Goldwyn production. This ambivalent critical attitude not only signals a profound cultural and linguistic divide, but also *Wuthering Heights*' status of film classic.

Both film classics, the Buñuel and the Wyler one, represent, from a translator's point of view, a 'transfer in excess of the original' or, in the terminology of George Steiner, a *betrayal by augment*. At the level of the Third MC and of the BTs, what are the consequences, in the Wyler's version of Emily Brontë's *Wuthering Heights*, of this 'transfer in excess'? To answer this question, I propose to take a close look at the remarkable sequence, in excess of the novel, where Heathcliff, prompted by Isabella's invitation to the county ball, makes his first public appearance at the Grange.[235] Finally, understanding how an Initiatory Love Journey transformed into a 'fatal love affair' becomes easier when remembering, then analysing the episode that compacts Cathy's forbidden reunion with Heathcliff, and her ensuing redemptive death.

* * *

With the images of the novel in mind, one might have the vague and unpleasant sensation that **the night sequence at the Grange** has been freshly imported from the film adaptation of a Jane Austen novel. But this impression does not generally linger. As the sequence unfolds, one realises that it complements, both in the original script and in the Goldwyn motion picture, the earlier, life-changing discovery by the then youngsters Cathy/Merle Oberon and Heathcliff/Laurence Olivier of the music of the party, the swish of the expensive dresses and the shimmer of the crystal chandeliers. At the time, Cathy/Merle could only grasp these marvels through the large panes of the Lintons' ballroom windows.

The country squire Heathcliff/Olivier, just as the magnificent Alejandro of the Buñuel picture, comes across as a congruent character to embody the main intra-diegetic outsider. Early on in the movie, in the fabricated episode of his severe bout of self-harm in the stables, which can easily remind the novel's readers of the wound inflicted by Lockwood on the ghost-child Cathy, his character would have merged with that of the reluctant participant in the story, this same old Lockwood, the epitomy of the outsider. Consequently, in the cardinal scenes of discovery, and the night sequence at the Grange inside (as well as outside) of the ballroom is certainly one of them, the film spectators have the option of identifying with Heathcliff/Olivier, rather than with the passive Mrs Dean (Flora Robson) who is Goldwyn's official reteller. The bland old traveller Lockwood of the 1939 production is no match for him. With the notable exception of his cautious visit to the Heights, and of his disturbing night there, when he confronts an unwelcoming clan of Yorkshire natives and fights the Dream Vision of Cathy, still eluding Heathcliff, he sits down between Mrs Ellen Dean and Dr Kenneth, then does not move away any longer from the cosiness of the fireplace.

It ensues that in the night sequence where Heathcliff/Olivier immerses himself into the social life of the Grange, the spectators are invited to follow him closely. Toland's camera, for most of the indoor and outdoor ballroom scenes, is thus either following in Olivier's footsteps (and showing the evening through his eyes) or operating behind him, at a safe distance from the bustle of the ballroom, in the surrounding darkness of the animated bushes, as though it were his own shadow. Olivier makes his stage entrance at the Grange as one could imagine him making his entrance on the British stage of the Old Vic, quite dashing and full of expectations, since he has come to win Cathy back. In the illuminated hallway, Toland's camera shoots the well-behaved children rehearsing their tidy version of the grown-ups' minuet, while Olivier introduces the spectators to 'a new territory' that is Hyper-Civilised and ruled by the Laws of Decorum. An initiatory movement going from the wild to the civilised runs contrary to the journey experienced by the novel's intra-diegetic outsiders and their mythical forerunners. This reversal of sense turns the First BT (Trespassing-Learning 1) upside down while the substituted surroundings, 'posh and hostile' rather than 'new and hostile', also become antithetical to Nature, and to the rule of the Present Moment.

In this context, the spectators' feelings of empathy are temporarily re-directed towards Isabella/Geraldine Fitzgerald; the equilibrium between social constraints and personal development is achieved, in this ballroom scene, by her, not by Olivier. She not only had the courage to ride on her own to the Heights and invite him in person to the Grange, but also now has the audacity to pair up and dance with him under the disapproving eyes of her brother Edgar (David Niven) and sister-in-law, who is thoroughly annoyed; as to the good Dr Kenneth, he is mystified! And, if this were not enough, Fitzgerald finally walks to and fetches Olivier from the balcony, where he has met with a reticent Oberon. While Olivier is supplanted by Fitzgerald in winsome spirit, the Second BT (Trespassing-Learning 2) cannot really apply to him either, and it can even less apply to Oberon who would not, until she lies on her deathbed, come to terms with the return of her childhood love as a rich man. In her performance on the balcony, Oberon does appear to be physically aware of the proximity of the moors, and of the pull of the Present Moment. However, her mixed feelings and emotions, which are silently expressed through all the touching hesitations in her body language, disappear behind the fluency of her self-defensive discourse, 'I am not the Cathy that was. [...] I'm somebody else ... I'm another man's wife and he loves me – and I love him'.[236] With Oberon, the scales are thus tilted towards Morality and Restraint, leaving Hyper-Morality and Passion (BT2, Interdict 2) literally out of the talking picture. Still, for some brief unspoken seconds, one feels that the dictates of Society are being totally eclipsed by the soft moon light that Toland has subtly re-created for Emily Brontë's silent lovers.

John Harrington comments upon this ambivalence within Wyler's *mise en scène* – and Oberon's interpretation – in his fine essay entitled 'Wyler as Auteur' (1981):

The interplay of *mise-en-scène* and moving camera continues as Heathcliff comes to the Lintons' party. [...] She is in control in the bright house, but she must struggle to maintain control as she enters the darkness of the balcony and feels the wind [...] Throughout the scene, the camera watches the two from the bushes, maintaining the camera's identity as a natural force; as Heathcliff talks with Isabella at the end of the scene, the camera pulls back into the trees and resumes its wait outside as Heathcliff reenters the bright and civilized world of the Lintons.[237]

* * *

Decorum and its demands are so potent in the minds of the inhabitants of the Grange that Oberon and Olivier are only given the smallest amount of time, alone, when they meet for their last encounter in order to be spiritually reunited. Oberon, very ill, finally manages to voice her feelings but her death itself, which belongs to the same sequence, asserts the ascendancy of Restraint and Morality. Ellen, Dr Kenneth and Edgar, the husband, in strict accordance with the Hays Code of the *Motion Pictures' Producers and Distributors Association* (whose main mission was the protection of the institution of marriage), are chaperoning the lovers' last private moments. One sequence earlier, when Oberon had set foot at the Heights to dissuade him from marrying Fitzgerald, Olivier deciphered and derided the hold that this stifling piece of bourgeois hypocrisy had on his lover. He would subsume it in a bitter-lame declaration to her, since all he then seemed able to promise her was a conventional extra-marital affair where she would have been left, on her own, in the throes of her Christian conscience:

> If your heart were only stronger than your dull fear of your God and the world, I would live silent and contented in your shadow ... Cathy! (*He moves to her – but she backs away.*) But no ... you must destroy me with that weakness you call virtue. (*He takes her arm.*) You must keep me tormented with that cruelty you think so pious.[238] [Original emphasis]

If Oberon's death still connects with the pre-Christian Gods, Eros and Thanatos, and cannot be washed from a serious hint of 'virtual' infidelity, it has been purged of any implication of illicit sex, and of licit pregnancy. All this combines to give the Indian-born, mixed-race and incredibly complex British actress Merle Oberon the simple stature of a photogenic Hollywood-born tragedian. From a mythocritical viewpoint, this would also have conspired to void the Third BT (BT3), had not Wyler fiercely resisted his producer's aesthetical directives for her death scene and re-injected the poignant density of Emily Brontë's tale to it:

> He [Goldwyn] wanted the director to redo her [Oberon] deathbed scene, reminding him, 'This is a somber scene and if you remember, I especially wanted Oberon beautifully gowned and beautifully photographed to help lighten it'. Wyler could not have disagreed

Figure 19. Merle Oberon (*Wuthering Heights* Poster, April 1939), *Kinematograph Weekly*

more. He [Wyler] believed that 'when beautiful movie stars allow themselves to look terrible, people think they are really acting'. Instead of cutting to gorgeous close-ups of Merle Oberon, Wyler and Mandell kept her in less glamorous longer shots as much as possible. After Goldwyn saw the death scene assembled, he wrote to the actress, congratulating her 'on the finest scene you have done since you have been in pictures. [...] I believe you should kiss Wyler for his direction of this scene'.[239]

Despite the quality of the composition (and editing) of this cathartic scene, it is unlikely that Merle Oberon, after their promising work on *These Three* (1936), would have forgiven so quickly, and kissed on the cheek so heartily, William Wyler. Throughout the *Wuthering Heights* film project, the latter made no secret that he would have preferred to cast Vivien Leigh and direct her in the leading role of Cathy – rather than in the secondary part of Isabella envisaged by Goldwyn – had not the producer staunchly supported Oberon, his favourite, beautiful contracted actress. Fortunately for Oberon, Leigh, who had actively anticipated to be chosen as the Scarlett O'Hara of the Selznick production, could not wait any longer to be reunited with Olivier in Los Angeles and, shortly after she arrived in America, secured her unforgettable part in *Gone with the Wind*.

For an audience of cinéphiles, the 1939 *Wuthering Heights* is a Hollywood classic with the same aura of fame as *Gone with the Wind*, which was released in the same year and robbed it of the Academy Award for Best Picture.[240] Ben Hecht, the highly esteemed screenwriter, worked on both scripts, and both films were rewarded for their cinematography: *Gone with the Wind* received the Academy Award for Best Cinematography in a colour movie, and *Wuthering Heights* in a black-and-white movie. Since then, the name of Gregg Toland has clung to *Wuthering Heights* as much as the names of William Wyler, Merle Oberon, Laurence Olivier or Samuel Goldwyn – and his renown, in all quarters, as an expert cinematographer has kept reflecting favourably on the production. In 1998, *Wuthering Heights* featured on the list of the *American Film Institute 100 Years...100 Movies* audience poll at place 73rd and, four years later, was ranked 15th in their twin survey – *AFI's 100 Years...100 Passions* – which concentrated on love stories only. Nonetheless, despite these excellent gradings and prestigious awards gathered across the Atlantic, the film has always been subjected to some contradictory appraisals in Great Britain and elsewhere:

[Holly] 'I read that story twice ... It doesn't *mean* anything.'
'Give me an example,' [Fred] said quietly, 'Of something that means something. In your opinion.'
'*Wuthering Heights*,' she [Holly] said, without hesitation. ...
[Fred] 'But that's unreasonable. You're talking about a work of genius.'
[Holly] 'It was wasn't it? *My wild sweet Cathy*. God, I cried buckets. I saw it ten times.'
[Fred] said 'Oh' with recognizable relief, 'oh' with shameful, rising inflection, 'the movie.'

Truman Capote, *Breakfast at Tiffany's* (1958)[241]

184

Chapter 23

Audience Response in the United Kingdom (1939-1978): An Insight into the Hermeneutic of *'Re-Appropriation'*

Before the triumph of the premières in New York and Hollywood, on 13th Apr, 1939, *Wuthering Heights* had represented an insurmountable marketing hurdle for Jock Lawrence's advertising team, at Samuel Goldwyn Productions. The team even got away with branding it, '*The strangest love story ever told*', since they could not really believe that a Hollywood-made movie populated by a British cast of unknowns, and showcasing a class-related tragedy set in stormy Yorkshire, had any chance of success:

> Still Goldwyn had a marketing problem on his hands. The salesmen were having difficulty booking so unconventional a film. The low marquee value of the cast did not help.[242]

After its box-office sensation on release in North America, some professional British reviewers, protective towards the interests of the national cinema exhibitors, were continuing to question the commercial value that a picture 'so sombre and melancholy' would have in Great Britain. Still, they could not help expatiating on its 'title and star values', which were *very* 'British'. *Wuthering Heights* could boast of the *high* 'marquee value' of its prestigious British cast, and would now be seen on all the cinema screens in the United Kingdom. *Wuthering Heights* would come back and play, as it were, at home:

> Whether the average audience will be willing to accept an entertainment so sombre and melancholy in key is problematical, but to discount doubt there are title and star values. These should look after the crowd [...] Its hold may not be so strong on the masses as on the more intelligent audiences, but it is nevertheless there. [243]

Fourteen years later, this 'sombre' and even 'morbid' quality was still fantasised by some as being dissuasively present for a more conservative category of television viewers, which the text for the 1953 BBC Audience Research Report well illustrates. Still, the actual audience figures for the play manifested a general attraction to tragedy and, by association, to its subdued mythical components:

> There was some adverse comment on the theme of *Wuthering Heights* from viewers who think that Sunday evening should be devoted to light comedy or 'popular' drama rather than anything 'morbid'.[244]

Another category of reviewers based in Québec, for instance, were impervious to tragedy, good acting and the subtle aesthetics of 'mental chiaroscuro'. They argued that the Goldwyn motion picture was setting bad moral standards owing to Cathy's, and quite subsidiarily Heathcliff's, unchristian motives:

> *Wuthering Heights*, scheduled for opening at Orpheum Friday (7), was banned by the Quebec Board of Censors from exhibition in this city and province on grounds of the 'infidelity' situation at the close of the picture.[245]

In 1968, the Hays Code definitely ceased to rule the Motion Picture Industry and, in the United States of America as in many other countries, there was a relaxing in censorship laws. Considering *Wuthering Heights*' staying power and international success, these reservations about the morality of the Goldwyn picture might appear somewhat antiquated. Nonetheless, they have come into play regularly since then, under different guises, and came into play forcefully when the early BBC television plays (1953 and 1962) were conceived: there, the domestic violence met by Isabella, and her spiralling down into alcoholism, were far more acceptable than the dark heroes' so-called 'unchristian motives'.

Upon its release in London, *Wuthering Heights* met with a newer, 'cross-heritage' type of criticism, which would then hit the Buñuel film when it was premièred in the United Kingdom back in 1984, to increase in intensity for the 'Paramount British Pictures' version (featuring the French actress Juliette Binoche) as the shooting was about to start there, on location, seven years later:

> Buñuel manages to introduce a few characteristic touches, but by any standards this film is well below his best. More importantly, it is not good Emily Brontë and as the earphones provided at the National Film Theatre reveal (in the absence of sub-titles) no use has been made of her great prose.[246] (Brontë Society Transactions, 1984)

It is nonetheless in the tongue-in-cheek article written by British film critic Seton Margrave, 'Sam's Tale from Brontë', and published in the *The Daily Mail* on 28th Apr, 1939 that the 'first' piece of cross-heritage criticism appeared.[247] His piece wavers between a strong sentiment of patriotic nostalgia and a stifled admiration for the box-office value of the Goldwyn production. At a deeper level, Margrave questioned the relevance of the cultural displacement that the English literary classic, *Wuthering Heights*, had undergone:

> [...] *Wuthering Heights* was filmed nearly 20 years ago at the old Ideal Studios at Elstree, and filmed very well with Milton Rosmer as the unhappy Heathcliffe [sic]. All this swank about Hollywood being courageous is all poppycock. There is nothing in English literature being filmed in Hollywood today that our old-time producers did not film – not so expensively but quite as sincerely – years and years ago.[248]

Margrave, while comparing the Hollywood picture to the earlier, and more legitimate (because indigenous) version made by Bramble and Stannard, gives a tangible dimension to the silent picture, which, in 1939, was still in the eyes and memories of a previous generation of picture-goers. Yet, the Goldwyn production completed the total eclipse of an old silent film that had long been defunct on the cinema screens, since virtually no silent films were being shown within the commercial network of exhibitors.[249]

For decades, the more superficial readings of the Wyler movie would overshadow the unconscious of the novel and, up until the 1978 *BBC Classic Serial*, these often were the only references for the television viewers. The predominance in their minds of the 1939 *Wuthering Heights* materialises with the response of the 1967 *Classic Serial*'s producer, David Conroy, to the genuine letter of interest by Miss Louise English (Gartree High School, Brocks Hill, Oadby, Leicestershire). The latter not only comes across as an attentive reader of Emily Brontë's novel, but also as an assiduous spectator of the serial, who looks forward to a repeat of the programme on BBC1. Both her letter, which is treated as a letter of complaint, and the answer to her letter attest to the sway of the Hollywood production on all UK audiences, be they composed of either 'profane' or 'professional' viewers. Conroy's lines, in particular, lay bare his own perception of the Goldwyn production, and reveal that the choice of his creative team to bypass the novel's three-dimensional architecture of *Mise en Abîme* was deliberate, a weaker option that would be taken up again, in 2008-2009, by the scenarist Peter Bowker:

> Obviously everyone has their own interpretation of a novel, but I tend to feel that the film over-romanticized the story and left out the passion. I like to think that in our dramatisation these elements were present, which made it truer to the novel. At the very early stage in the planning of this dramatisation we decided that the story would best be served by telling it in chronological order, and so although Lockwood did not appear in the same place as in the novel he most definitely appears towards the end of the serial.
> At the moment there are no plans for repeating this programme on BBC-1 [...]
> (David Conroy, 22nd Nov, 1967)

The ongoing phenomenon of acclimatisation experienced by the Wyler movie, in the United Kingdom, until the 1992 production by Paramount British Pictures took over part of the cultural territory occupied by the well-remembered Hollywood film classic, can be distinctly spotted in a comment excerpted from the Audience Report for the 1978 serial (Week 39). This comment asserts, forty years after the initial release, its unrelenting *appropriation* by the British public at large:

> It was clear that the Olivier film version was for some the definitive adaptation, and tonight's episode was compared unfavourably with it, particularly in terms of overall

atmosphere. A number of reporting viewers considered this television version to be 'too miserable, too depressing and too gruesome'.[250]

* * *

The villainous Cathy created by Hecht and MacArthur, with her mercenary romanticism (and, in the words of George Bluestone, 'strong mythopoeic tendencies'), was transported into the 1967 serial and early television plays.[251] It is through her characterisation that the BBC television productions, with the notable exception of the just cited 1978 serial, appear to be most susceptible to the influence of the 'Olivier film version'. One of Balthus' most famous Persian-ink drawings, which belongs to his *Wuthering Heights* series, is entitled 'La toilette de Cathy' (1935). To advance our critical journey, why not superimpose this emblematic drawing on to a key scene of the 1939 film classic where Merle Oberon is at her toilette?

Heathcliff/Olivier, who has guessed that Cathy/Oberon is getting ready to welcome her admirer from Thrushcross Grange, faces her provocative comments. Those are hurled at him in response to his prickly questioning about the uneven amount of time she divides between her suitor, David Niven, and himself. Fundamentally, the protagonists' row originates in their last race to Penistone Crags, the childhood 'Castle' that they must now renounce to as grown-ups. From Cathy/Oberon's point of view, Olivier has failed to keep his promise of sailing away to seek his fortune. She ends up calling him 'a stable boy', and the argument spirals out of control: 'You had your chance to be something else than a beggar beside a road', declares the Hollywood Studios heroine, speaking the words created by Hecht and MacArthur. The dark hero of the 1939 production loses his nerve, and 'un-Brontëanly' slaps her hard on both cheeks.

Follows Olivier's full-on caricature of the character interpreted by Oberon, a rather vile portrait of her that would stick to the subsequent BBC Cathys:

Yes – yes, tell the dirty stable boy to let go of you – he soils your pretty dress. But who soils your heart? Who turns you into a vain, cheap, worldly fool … Linton does! You'll let yourself be loved by him because it pleases your stupid, greedy vanity […][252]

This is a far cry from Balthus' tenderly narcissistic, yet erotic, representations of Cathy as she is admiring herself in the mirror. The Third MC (MC3) is completely transformed, not in the camerawork, but at the level of the dialogues and original screenplay. This radical transformation constitutes, from a translator's viewpoint, a 'failing' that, in the words of George Steiner, 'projects as on to a screen, a resistant vitality', one of 'the opaque centres of specific genius in the original' text.[253] Un-Brontëan and un-Balthusian, Oberon at her toilette, busy at doing her hair and touching up her dress, is the epitome of a pervasive *betrayal by augment*.

Finally, the transfiguration of Cathy's character is manifest not only in the re-interpreted scene of Cathy at her toilette, but also in the cardinal sequences of the county ball and of her redemptive death, which we have discussed previously. This transfiguration of hers occurs much earlier in the film when the children actors are seen to be racing to, then enjoying Penistone Crags. Their first acquaintance with the liminal site is marked by a communion with Nature that makes the First MC (MC1) very much tangible. Nonetheless, the feeling of intense immediacy – the 'light weight of the Present Moment', as it were – which is related by the images on the screen, is negated by the non-performative, mythopoeic speech that the child Cathy is made to utter:

> Your father was Emperor of China and your mother an Indian Queen. Your were kidnapped by wicked sailors and brought to England. But I'm glad they did it, because I've always wanted to know somebody of noble birth instead of … vulgar little peasants like Hindley.[254]

Little Cathy's words are anchored into a romanticised past and a dreamt future. They are borrowed from Ellen's comforting words to the child Heathcliff in Chapter VII, 'You're fit for a prince in disguise'. These words are spoken by Ellen, in situation, after she has managed to 'dress him smart' for the Christmas party that takes place at the Heights, and put an end to a dark, intense bout of sulking.[255] Therefore, little Cathy's words, as spoken by Hecht and MacArthur, have been ripped from their context in order to start beating into shape the mercenary personality of their Cathy. Pamela Mills, in her often-quoted essay, 'Wyler's Version of Brontë's Storms in *Wuthering Heights*' (1996), aims to inventory the film's numerous departures from the situations and characters of the novel and, in doing so, flags up Cathy's 'flawed' cultural translation:

> The major flaw of Goldwyn's *Wuthering Heights* is the portrayal of Catherine's motives. In the film, she is drawn to the glamour of Edgar's way of life for vain and other obvious reasons, making her somewhat a villain for her treatment of Heathcliff […].[256]

Mills' approach, which is mainly inventorial, fails at explaining those departures that she rarely, if ever, considers as flaws in the dynamic structures of story-retelling. All the departures from the text, regardless of their impact on the audiences, are justified by the free *appropriative* use of Emily Brontë's novel that Mills grants, on her own initiative, to her preferred 'auteurist' version. With such auteurist tactics, the hermeneutic motion, which, in the ideal translative process, renders translators as well as music and film adaptators accountable for their interpretations, adjustments and hesitations, comes short of the phase of *compensation* that restores the balance between source and translated texts. Taxonomic methods based on auteurism cannot replace the mapping out of 'some reasoned descriptions of processes',[257] which George Steiner recommends for enriching the vision of translators and critics alike:

Genuine translation will, therefore, seek to equalise, though the mediating steps may be lengthy and oblique. [...] The ideal, never accomplished, is one of total counterpart or re-petition – an asking again – which is not, however, a tautology. No such perfect 'double' exists. But the ideal makes explicit the demand for equity in the hermeneutic process.[258]

The BBC Teleplays

This section makes use of records from the BBC Written Archives Centre in Caversham, Reading

Archival References

1948/1953 Plays:
BBCWAC T5/604/1 'Wuthering Heights' 1948-1953
BBC Audience Research Report (AR) on Wuthering Heights, tx. Sunday, 6th Dec, 1953 (Week 50) – 8.40-10.35 pm – R9/7/9

1962 Programme:
BBCWAC T48/357/1 Nigel Kneale (TV Script Unit File)
BBCWAC T5/2440/1 'Wuthering Heights' 1962
BBC Audience Research Report (AR) on Wuthering Heights, tx. Friday, 11th May, 1962 (Week 19) – 9.25-11.00 pm – R9/7/58

In the three early television plays produced by the BBC, the literal translation of the superficial structures of the 1939 picture – which are easily accessible, as we have just seen, through the dialogues and screenplay – led to an inadequate rendition of the First MC (MC1; Correspondence between Settings and Characters) and to a caricatural treatment of the Third (MC3; Correspondence between Love and the Sacred), transforming them into soap operas. More specifically, this literal translation of an influential cinematic model affected the standing of the main female characters in all three televised dramas, and thus impacted adversely not only on the character of Cathy but also on the character of Isabella, who is given the prominence of an anti-heroine in the Wyler version. As a result, the Isabellas of the BBC teleplays, and this includes the lovely June Thorburn, who had just impersonated Kitty in *Anna Karenina* – another high-profile BBC-commissioned adaptation (1961) – would end up having little in common with the dignified Isabella created by Geraldine Fitzgerald. The balance of characters present in the script of Hecht and MacArthur, and achieved in the film through the combined performances of Oberon, Olivier, Niven and Fitzgerald, would be profoundly unsettled:

[…] the de-emphasis of the Lintons' weakness, in the face of two dynamic characters, is an acceptable alteration. […] He [Nugent] described their [David Niven's and Geraldine Fitzgerald's] performances as 'dignified and poignant characterizations […]'[259]

Isabella's character, in those BBC teleplays, was not offset by Wyler's intelligent direction of Olivier, whose Shakespearean persona was in itself a defence against an excess of casual misogyny. It did not benefit either from the splendid cinema sets designed by James Basevi. These sets and their association to performance and meaning, in the context of the translation of *Wuthering Heights* on to a television (or cinema) screen, are of major importance since they guarantee the integrity of MC1. Their inadequacy, as in the early television play of 1948, entails the disqualification of the cultural translation to be put on an equal footing with Emily Brontë's novel. Some translative schemata that flout both main MCs and main female characters were overtly applied by Suri Krishnamma to his MTV version of *Wuthering Heights* (2003), which thus appears both trashly sexualised and fashionably sanitised. Some other schemata, similarly faulty, were astutely turned to their best parodic advantage by the troupe of the Monty Pythons, when it created its 'Semaphore Version of *Wuthering Heights*' (July 1970).[260] In that hilarious sketch, Heathcliff and (of necessity, a cross-dressed) Cathy were made to stand on top of two towering hills that seemed to be so despairingly far apart that the Rapunzel-like captives had to resort to basic semaphores in order to communicate! This instantaneously severs the ineffable passion they feel for one another from its dynamic structures and mythical components, and empties it of all substance.

The same severing of the story from its unconscious inadvertently occurred to the first teleplay due to its bungled copy of the televisually un-replicable 'Olivier' version, and to the lack of self-assurance of the new medium. Back in 1948, Robert McDermot was at the head of the BBC Drama Department. On 8th March, he would address a scathing memo to the esteemed producer and director, George More O'Ferrall. In this memo, dug out again from the BBC Written Archives, McDermot writes that *Wuthering Heights* was an unusually bad teleplay, with some particularly badly written dialogues (John Davison and Alfred Sangster). Nonetheless, even after having taken the live televised play to bits, while paying heed not to offend O'Ferrall directly, the Head of Drama cannot get himself to say that the Studio A of Alexandra Palace, used for setting up the main hall of Wuthering Heights (Barry Learoyd) and staging in the exact same fashion the two antithetical families of the Earnshaws and the Lintons (as well as the two generations, without giving the audience any other tangible indication of time passing than with three captions), was the one mistake too many that had finished transforming the first teleplay of *Wuthering Heights* into a bad parody.[261]

⋆   ⋆   ⋆

Five years later, in 1953, the year when the first of his ground-breaking science-fiction serials, *The Quatermass Experiment*, was transmitted live, Nigel Kneale chose to simplify the structure of the Hollywood scenario of *Wuthering Heights*. The writer, who was one of the most experienced of an early generation of television playwrights, became immediately concerned about the inescapable 'technical difficulties' then attached to live performance, and swiftly decided to erase the second generation featuring in the 1948 screenplay as well as the childhood scenes typifying the beginning of the Hollywood film. He re-introduced the character of Lockwood, to whom he gave much more visibility than even Hecht and MacArthur had dared to lend him. The only hitch was that, in the absence of Hareton Earnshaw, Catherine Linton and Linton Heathcliff, the character of Isabella emerging from this new live televised combination was left to bear, alone, the brunt of the revenge carried out by Heathcliff:

> [...] the inevitable compression would bring the new generations popping up like Jack-in-the boxes, with babies shooting to manhood in a few minutes' playing time. Add the technical difficulty of 'ageing' characters successfully during a television performance. [...]

> The second generation – Hareton Earnshaw, Catherine Linton, and Linton Heathcliff – would not appear but since their part in the plot is mainly to show, as victims, the relentless working-out of Heathcliff's ambitions, it could be taken over by the unhappy Isabella. This broken wife of Heathcliff is therefore kept alive as one of the characters encountered at Wuthering Heights twenty years later, when the genial Mr. Lockwood calls on his new landlord and happens upon the whole story – for I have retained the dramatic narration-framework of the book.[262]

Despite Kneale's careful retention of the 'dramatic narration-framework' through the accentuated presence of Lockwood, the diegetic framework of the novel had been completely revisited, weakened by the suppression of the characters 'en devenir'. Kneale's simplified screenplay, by allowing the learning trajectories to vanish and Isabella to assume, in the context of a British society still dominated by sharp but unspoken gender inequalities, the role of a half-consenting victim, made her character lose its nobility of feelings, as well as its alluring distance from the viewers. Even the fresher 1962 production with its first-class crew, A.A. Englander on his Mole-Richardson camera crane, and glittering cast, Claire Bloom (the 'Helena' of Osborne, Richardson and Kneale's *Look Back in Anger* (1959) and the 'Anna Karenina' of the sophisticated 1961 teleplay by Cartier) as Cathy, Keith Michell as Heathcliff, June Thorburn (the Kitty of the previously cited teleplay) as Isabella and, finally, David McCallum as Edgar, did not succeed in attenuating the misogynistic and parodic feel inherited from the screenplay that Kneale had hastily conceived nine years before.

Within the expanded confines of White City's BBC TC3 studio stage, Isabella remained a small-minded and thoroughly maltreated wife, who could hardly command, especially on a weekend night, the empathy of a disengaged and forgetful television audience. Most of the viewers turned a blind eye to June Thorburn's domestic plight, and to what must have appeared, to a few spectators only, as her husband's attempt to strangle her when the pre-recorded drama, which still bore some characteristics of a live televised performance, was broadcast in 1962. (It is possible for the twenty-first-century television fans and critics to view, at the BFI, this archived programme, in its five reels of 35-mm telerecorded footage, and easily spot the 'strangulation scene'.) This shocking aspect of the teleplays, damaging not only Isabella's standing in the story but also the formidable aspects of Love and the Sacred (MC3), was obliterated in all the BBC contemporary appraisals I came across, while being seriously investigated by New Scotland Yard in May-June 1962, following the transmission of the programme on Friday 11th May, between 2126.15 and 2259.42:

> In my capacity as an officer of the British Broadcasting Corporation, I have been asked by Police for details of any *strangulation scene* that was shown on B.B.C Television on Friday 11th May, 1962, and for a week prior to this date.
> [...] In this play there is a *strangulation scene* which showed the character 'Heathcliff' *grasping 'Isabella' round the throat.*[263] [Emphasis added]

In any event, the Scotland Yard inquest neither prevented the repeat of the programme eleven months later on the BBC, nor its sale to the USA station, WOR, 'for two showings over three years to October 1965 app'!

*   *   *

While Rudolph Cartier was still in the early stages of reviving their 1953 production, Kneale would give a very honest reply (dated 31st Jan, 1962) to Robin Wade, then Senior Assistant in the BBC Script Department. Wade was diplomatically contacting Kneale, in order to gauge his interest for this new *Wuthering Heights* television venture, for which the BBC was unashamedly recycling his old screenplay. By his own avowal, in this response to Wade's letter, Kneale admitted that he had somehow managed to complete the writing of his 1953 piece in 'eight days flat'. The Senior Assistant had guessed it right, the much sought-after playwright, this time round, would not be dragged into this *Wuthering Heights* business: Kneale did not even want his name to appear in the credits![264] One can thus legitimately infer that Kneale dreaded the parodic flavour of early soap opera that his script was yet again likely to have on the small screen, even in a much more modern teleplay that had been telerecorded in order to be transported to another television territory, and could now boast of a powerful wind machine, of an improved wardrobe and musical soundtrack, and of this amazing-looking counterbalanced camera crane – the Mole-Richardson.

Lockwood could neither gain much wisdom nor much inspiration from the domestic war that had inexplicably swept away the two antithetical families. He became a stranger who is immersed into a dismal domestic drama, not an intra-diegetic outsider progressively drawn into an Initiatory Love Journey. And, although Kneale had envisaged staying closer to Emily Brontë's final heathen tableau,[265] the Lockwood of both his teleplays, looking at the headstones, would conclude with this lame epigram, 'They rest from their labours' (*Revelation, Apocalypse* 14:13), awkwardly plucked as it seems from the Anglican hymn *For All The Saints*, an epigram that did not leave any hope of a connection between the Living and the Dead (MC2), between the self-righteous narrator and such an unprincipled pair of sinners as Cathy and Heathcliff! Still, according to the BBC Audience Research Reports for the 1953 and 1962 productions, both teleplays would be well received by a public who was then rather forgiving when it came to poor lighting and coarse special effects:

> The technique used in presenting the play was generally commended. [...] It was often felt too that stronger lighting in some of the gloomier scenes would not have come amiss.
>
> (1953)

> [...] viewers found fault with minor details (as, for instance, snow that 'looked like crystalline globules' instead of flakes).
>
> (1962)

The production, in 1953, was described as being 'intensely dramatic and gripping' and, by 1962, the comments printed in the BBC Audience Report had become increasingly elaborate and contrived:

> [...] they [some viewers] were watching a 'very gripping and moving story' that while touching 'certain deep chords of human feeling' had also the power to enchant by 'its magic and romance'.
>
> (1962)

The popular success of the 1962 production was quite real, albeit totally baffling. Therefore, even though Nigel Kneale had eloquently opted out from the credits, Wade had been well inspired to insist on keeping him on the payroll for the revived teleplay. On 30th May, 1962, Donald Baverstock, BBC Assistant Controller of Programmes, would be dictating a memo, which retrospectively is fairly amusing, in a desperate attempt to make sense of the sky-high audience ratings of a fairly uneven version of *Wuthering Heights*: the members of the Programme Planning Committee had not succeeded in reaching a satisfactory conclusion![266] Baverstock could have addressed this rare memo to Wade, the Script Department Senior Assistant. He could also have addressed it to Donald Wilson, Head of the Script Department. Wilson would have overseen the work on all adaptations be they treated as series, serials or plays until the newly appointed Head of Drama, Sydney Newman, reshaped the entire department and had him focus mostly on serials and *Dr Who*.

The Lindsay Anderson Film Project (1963-1965)

A little more time and much more work and passion would be needed to conjure up, on television, a (cultural) translation of *Wuthering Heights* that trusts the serial format and seeks to establish a dialogue with the unconscious of both the source text (Emily Brontë's novel) and the seminal version of 1939 (Wyler's film). In Yorkshire, a propitious atmosphere and an open, flexible frame of mind were crystallising to partake in a creative balance finally reached with the 1967 BBC2 production; those favourable signs were particularly well captured by Peter Holdsworth, the *Telegraph and Argus* show business chronicler, in his article 'Another *Wuthering Heights* Film: Location Work in West Riding will Start Next Year' (2nd July, 1964), which will be cited next. This interesting article, and all the primary-source material around which I am evolving this section, arise from the Lindsay Anderson Collection, 'Archives and Special Collections' of the University Library of Stirling.

It so happened that two practitioners of the Free Cinema movement, director Lindsay Anderson and (screen)writer David Storey, had decided on tackling the impossible adaptation, while very much hoping to cast in the title role for their *Heathcliff* (or *Love For Life*) a charismatic actor with whom they had previously worked, Richard Harris. In their first feature film, *This Sporting Life* (1963), his compelling interpretation of rugby league player, Frank Machin, had won Harris the Palme d'Or for Best Actor (Cannes Film Festival) and given Free Cinema instantaneous acclaim, especially abroad, in the *Nouvelle Vague* press. By the Spring of 1964, Harris would already have dedicated a great deal of his mental space to this new *Heathcliff* project, and comes across, through his correspondence, as a 'fin connaisseur' of Emily Brontë's novel since he did much more than respond superficially to the scripted ideas of David Storey. When Summer arrived, Holdsworth had every reason to be writing enthusiastically about this forthcoming adaptation. It was to supersede the 'old' Hollywood classic in creativity and critical insight:

> The turbulent awe-inspiring love between Heathcliff and Cathy is to blossom wildly again on the moors of Haworth. Among the gorse and bracken, amid the cold stone walls and windswept hills, one of the greatest romances will return to its birthplace. [...]
>
> This creative, power-packed art of film-making should ideally suit the dark, introspective mood of *Wuthering Heights*. Although the earlier film of the novel was a success it was made at a time when picture-making was far less advanced and it never reached the subconscious depths of the novel.[267]

The succès d'estime of the Free Cinema trilogy produced by Karel Reisz had culminated in Anderson's *This Sporting Life*; this gave the successful director enough credibility to go and look himself for financial backing. To push forward his new film project, not only did Anderson knock on the door of Julian Wintle – a member of the short-lived Bryanston Films, a consortium chaired by Michael Balcon (the adjuvant to young Hitchcock's film career who logically turned, thirty-three years later, supporter of the British New Wave) – but also went on to prompt his favourite male lead, who had made a huge impact in North America with his part in *The Guns of Navarone* (1961), to get in touch with Walter Reade's Continental Film Distributors. The aim of the manoeuvre was to secure the distribution of their version of *Wuthering Heights* in some of the most prestigious cinema theatres on the East Coast.

Notwithstanding, at a very early stage, and despite their best efforts, the production of the script of *Heathcliff* (or *Love For Life*) slipped away from the intended British-dominated partnership. In this respect, one can safely conjecture that the failure of Bryanston Films to reap the benefits of Tony Richardson's promising *Tom Jones* (1963) as they refused to produce it in colour, and their inevitable demise shortly after that fateful misjudgement, hit Anderson's *Heathcliff* project head-on.

In the letter, rather admonitory in tone (6th Sept, 1964), which Anderson wrote to Sandford Lieberson, his agent, then based in Los Angeles, there is no longer any mention of direct British funding, or even support, for the production.[268] The focus is on Anderson's difficulty to communicate effectively with Lieberson and 'Creative Management Associates' (CMA). This issue precipitated the 'remov[al] of the whole project' from ex-Gold Medal Studios' boss Martin Poll to patron of the arts Joseph Lebworth, both New Yorkers, while preventing Anderson from taking up the middle-ground option embodied by George ('Bud') Ornstein, production head of the European Division of United Artists.

What is vividly described is the lack of momentum for the whole of Anderson's portfolio on the West Coast as CMA played actor against director. By rendering impossible Harris' commitment to the staged adaptation of *Hamlet*, which Anderson had 'largely undertaken on his behalf' in England, CMA were also giving amunition to top-level United Artists executive David V. Picker, who kept postponing his promise to pay the director for his 'work on the production of the present [*Heathcliff*] script', in Hollywood.

From Harris' perspective, however, there were some practical hindrances as well as some interpretative resistances for not going ahead with the project, after an initial phase of profound involvement. Firstly, it becomes clear, with only one quick look at his schedule, that he was swamped with many proposals thrown at him at the same time. These were ranging from Michelangelo Antonioni's *Il deserto rosso* (*The Red Desert*, Italy) and Sam Peckinpah's *Major Dundee* (Mexico), both shot, in rapid succession, on foreign locations in the Spring of 1964, to Mauro Bolognini's *Gli amanti celebri* (Italy), the middle part of *I tre volti* (*The Three Faces*), and *The Heroes of Telemark* (Norway). The filming of Anthony Mann's war film took place in the late Summer of 1964 and over the wintry months of 1964-1965. Additionally, Harris was soon to be absorbed, without much time for a breather, into George Roy Hill's *Hawaii* (Los Angeles and Hawaii) and John Huston's *The Bible* (Italy), which finished filling up his 1965 diary.

Secondly, as far as *Wuthering Heights* and its translation into the David Storey's script was concerned, Harris had some obsessions, and legitimate reservations, especially about the ending. He convincingly argues in his 'Notes on *Heathcliff*', dated 2nd Apr, 1964, that the ending was 'a cheat' as it conveyed a distorted picture of both the title character and his 'love', Catherine, by ignoring the elemental dimension of their struggle.[269]

These extraordinary 'notes', a two-page document devoted to the character of Heathcliff rather than to the entirety of the script, was handed in to Anderson when he came to visit the actor in Mexico City, on the shoot of *Major Dundee*, in order to discuss their *Wuthering Heights* film project. Harris' full-blown notes make tangible the close intellectual partnership between actor and director, and attest to his high degree of involvement in the process of cultural translation, as he lays bare the unconscious of *Wuthering Heights*.

All First (Connection between Human Beings and awe-inspiring Natural forces or Beauties), Second (Connection between the Living and the Dead) and Third (passage from Destructive to Reconstructive Passion; Love and the Sacred) MCs are there. What is more, in his own style and words, Harris recognises the Second (Empathy with Heathcliff's Plight) and Third (Undergoing a Life-Changing Moral Transformation) BTs as he takes on, without consciously noticing it, the viewpoint of the Lockwood of the novel. By weaving his idiosyncratic metaphor of the 'stream', Harris relates to Anderson his reading of the antagonistic forces that inhabit landscapes and characters in Emily Brontë's novel. Only falling short of using the terms favoured by the pre-Raphaelites in their Aesthetics of Landscape, the *sublime* and the *beautiful*, but having read David Cecil, Harris understands these forces as being canalised into two distinct 'channels', and classifies the characters into two natural categories, the 'children of storm' and the 'children of calm'. He then assimilates the story line to a disruption, then painful restoration, of these original categories.

The actor stresses the importance of this symbolic reading, which, he contends, surpasses Storey's scripted rendition of Heathcliff as a mercenary villain, wreaking havoc around him out of spite and envy. Harris is thus very much attuned to the source novel where, for a momentary stasis heralded by Lockwood's comeback on to the moors, all natural forces, inclusive of the irreconcilable 'channels', are slowly converging together,[270] in a final heathen tableau that Nigel Kneale was never to see *mise en scène* in his early televised adaptations.

Harris, like Storey, leaves *Psyche*'s myth, and the self-revelatory journey experienced by the heroine, aside but, contrary to him, believes in the value of the recounting, for the extra-diegetic observers, of Heathcliff's hyper-moral stance and laborious invocation of Cathy. Heathcliff, underlines Harris, succeeds in conjuring up, before his eyes, his 'immortal love'.[271] The similarity is so striking between Harris' prose (1964), '[Catherine's magnetic power] claws so strongly that it breaks through the veil of mortality to manifest itself to the physical eye in the person of a ghost', and the quintessential 'conte cruel' by Villiers de l'Isle-Adam, *Véra* (1874), where the Count d'Athol/Heathcliff is said to have 'hollowed out in the air the shape of his love', and Véra/Cathy appears to be 'playing with the invisible, as a child might',[272] that both the Irish actor and French marquis seem to be literally possessed by the last chapters of *Wuthering Heights* (1847).

In the novel, the other intra-diegetic outsider, Nellie Dean, lets us witness through her own eyes how Heathcliff finally reaches Cathy/Véra after having slowly induced, over a period of eighteen years, his visions of her. At the end of his initiatory trial, Heathcliff *will* actually *see* her ('il la Véra/verra'):

> With a sweep of his hand, he cleared a vacant space in front among the breakfast things, and leant forward to gaze more at ease.
>
> Now, I perceived he was not looking at the wall, for when I regarded him alone, it seemed exactly, that he gazed at something within two yards distance. And, whatever it was, it communicated, apparently, both pleasure and pain, in exquisite extremes, at least, the anguished, yet raptured expression of his countenance suggested that idea.
>
> The fancied object was not fixed, either; his eyes pursued it with unwearied vigilance; and, even in speaking to me, were never weaned away.[273]

With such a heartfelt response to the Second MC (MC2), and the deep accents of Symbolism, we can fully grasp Harris' objection to the hellish retribution imagined by Storey. In his *Jane Eyre* finale, the *sublime* heroes perish in the flames of a Wuthering Heights burnt down by an angry mob of villagers punishing Heathcliff for his acquisitive lust over the land, the houses and the corpse of his beloved. This type of ending, argues Harris, would curtail the 'emotional participation' of the audience, as well as rob Heathcliff and Catherine of their regal standing in the story.

* * *

As early as 1964, as indicated by the dates appearing on the contact strips and negatives of photographs of that same Special Collection, Lindsay Anderson began scouting the West Riding part of Yorkshire for original locations and ideas. However, without Richard Harris in the role of Heathcliff, the film project was losing much creative force, and this loss was insurmountable. By the late Spring of the following year, the 'Free Cinema' film director would have had to shelve *Heathcliff* (or *Love For Life*).

Despite this failed attempt, the determination to tackle the classic was definitely in the air. The BBC was now ready to rise to the challenge of *Wuthering Heights*' adaptation by taking full advantage of telerecording, and actualising the format of the television serial.

The space of the former television plays was to open up on to outdoor locations, and the first *Wuthering Heights'* BBC serial would be characterised by a return to the spirit of the shooting on the moors typified by the Albert V. Bramble's lost picture.

The BBC2 Classic Serials

This section makes use of records from the BBC Written Archives Centre in Caversham, Reading

Archival References

1967 Programme:
BBCWAC T48/374/2 Hugh Leonard (TV Script Unit File)
BBCWAC T5/704/1 & T5/704/2 'Wuthering Heights' – Filming Files 1 & 2
BBCWAC T5/705/1 'Wuthering Heights' – General File
BBCWAC T5/706/1 'Wuthering Heights' – Ep. 1-4
BBCWAC T5/2541/1 'Wuthering Heights'
BBC Audience Research Report (AR) on 'Wuthering Heights', tx. Saturday, 18th Nov, 1967 (Week 47) – 7.55-8.40 pm
BBC Audience Research Report (AR) on 'Wuthering Heights', tx. Saturday, 28th Oct, 1967 (Week 44) – 7.55-8.40 pm

In comparison with the early BBC productions (the 1953 teleplay and its popular 1962 revival), the 1967 and 1978 BBC2 *Classic Serials*, which were given much more thought and resources, enjoyed a more divided audience response that did not reflect so much on Reaction Indexes, which could never have reached the top figure (79) of the Cartier-Kneale *Wuthering Heights* teleplays anyway, as in the critiques. The higher expectations of the television audiences, as well as the commencement of the standardisation of their tastes by the late 1970s, seem to have led to this new critical assertiveness that was becoming manifest, as we will soon see, in the recording of longer strings of original comments on to the BBC Audience Research forms. Starting in 1953, the 'BBC Audience Research staff' would be recording, on their dedicated forms, 'viewing ratings and qualitative reactions to BBC programmes based on an audience sample'.[274] On these reports, which have already substantiated and enlivened some earlier discussions in this chapter, the BBC Audience Research staff were often trying to rationalise their findings, and they ostensibly did so for the 1978 production and its first episode:

[…] as the Reaction Index would indicate [66], Wuthering Heights was felt by a substantial number to have got off to a fairly promising start. Those who said it had exercised a fair degree of appeal were slightly the majority and this section of the reporting audience tended to feel that the power of the original novel came through. Unfortunately, however, favourable [sic] impressed viewers were much less forthcoming about their reasons for enjoying tonight's programme than were those who had disparaging criticisms to make.

The 1967 and 1978 programme-makers strove to present convincing realistic settings, and this was picked up by even the less 'forthcoming' and more 'critical' viewers of the reporting 1978 audience. By contrast, the huge efforts of the earlier *Classic Serial* to actualise the First MC (MC1) through extensive location filming, complemented by the construction of transportable matching sets in studio, if they appear to have been rewarded by adequate Reaction Indexes (Episode 1: 71, Episode 2: 64, Episode 3: 67 and Episode 4: 70), were not actually perceived by all viewers. The 'setting of the sets', in particular, had been an intense source of worry for the *Classic Serial* team inasmuch as Production Designer, Peter Kindred, had to hop from Alexandra Palace (Episode 1) to Riverside in Hammersmith (Episode 2) in order to finally land in the TC3 studio heaven of Television Centre at White City (Episodes 3 and 4).[275] While the actualisation of MC1 was finely noticed by some in the BBC Audience Research sample who commented on the effectiveness of the 'moor scenes' as well as described, in their own words, the contrasting houses (Week 47 AR) with a special mention for the 'Earnshaw's farmhouse' that conveyed 'the harsh personalities of its inhabitants' (Week 44 AR), others in the sample failed to 'realise that the exterior shots had actually been filmed on the Yorkshire moors near Haworth'. Further, there were some 'unfavourable comparisons' made between the serial and the 'film version with Laurence Olivier as Heathcliff' (Week 47 AR) by those viewers reminiscing with nostalgia the spectacular cinematography of the Goldwyn production of 1939.

At all events, the *Classic Serials* corresponded to some truly innovative pieces of television craftmanship, pushing back the limits of the existing BBC television regulations to include child actors, adjusting to the need for post-production editing, as well as aiming to strike a balance, aesthetically, between filmed inserts and videotaped studio material. The location sequences shot on 16-mm film and, in particular, those outdoor scenes relying on horse-riding and dialogue were characterised by a Free Cinema 'dramatised documentary' feel. The single-camera film inserts (35-mm and 16-mm) had to be smoothly blended into the well-defined grain and 'notan' lighting of the multi-camera studio drama shot on videotape (VT) that allowed for a vivid televisual experience. As kindly retold by Jonathan Bignell in his illuminating article dedicated to the *Wednesday Play*:

Film pictures were not straightforwardly 'better' than studio video, since studio video cameras had higher definition pictures than 16-mm film cameras. While film appeared to connote realism when used in exterior locations, it had a softer picture quality. However, high definition video cameras could pick up details of a constructed set, or a costume, that directors might want to hide because they revealed the artifice of studio production. The electronic cameras used in the multi-camera studio also required high intensity of light in order to register satisfactory pictures, so a director's choice of chiaroscuro effects and high-contrast lighting in television visual design was, although possible, both time-consuming and risky.[276]

Technological, aesthetical and regulatory issues would have represented constant yet exciting challenges to rise to, especially back in 1967. The memo, which I freely entitled 'Finding Young Hareton' during my research at the BBC Written Archives (Caversham, Reading), emanates from the director of the 1967 serialised drama, Peter Sasdy. It was primarily addressed to David Conroy, the producer of the BBC2 *Classic Serials*, but was probably copied to Head of Drama, Sydney Newman. This memo dated 26th June spells out how tricky the casting, in Episode 3 ('The Abduction'), of a 'young Yorkshire boy' had been, from a legal point of view, for the BBC:

> I will be most grateful if you could get Head of Drama Group's agreement to use a young Yorkshire boy to play this part, which will involve one studio date as well as location filming, even though this is against the rules of the Corporation and the Greater London Council.[277]

Dated 9th Aug, a 'thank you' note addressed to John Rutherford Esq., Kildwick Hall, which is signed by Ian Strachan, production assistant to Sasdy, reveals that this child, chosen to incarnate Hareton on the television screens, was the son of the 'house', Timothy:

> The first episode goes out on BBC-2 on October 28th. Timothy features in Episode 3, on November 11th. He looked marvellous on the rushes so I hope you will be able to watch him on the screen.[278]

The participation of eighteen children of 'Haworth County Primary Junior & Infants's School' is attested by another 'thank you' note dated, again, 9th Aug, and supported by the detailed letter, including the name and height of each child, that was drafted by the headmaster in person, Mr Wilding, on 19th July. The latter was responding to the same Strachan, who would subsequently reiterate the promise of a cheque for the school fund, and pass on the good news of Timothy's cameo appearance to Kildwick Hall. The location filming for the 1967 dramatisation of *Wuthering Heights*, which took place at the height of Summer (between Wednesday 26th July and Thursday 3rd Aug), reflects the active involvement of a whole local community. This is strongly reminiscent of the positive frame of mind that greeted the preparation of the lost Bramble-Stannard picture, and of the sustained interest accompanying its filming, on the same locations, in the Spring of 1920.

Director Peter Sasdy and Producer David Conroy had a favourable impact on the translation of the First MC (MC1; Correspondence between Settings and Characters) on to the small screen. Sasdy, who was responsible, in great part, for the logistics of the serial, organised the London rehearsing facilities and recording studios and, pushed by his director, eventually managed to book him the cameraman, John Kealing, with whom he had worked efficiently on a long 'synced' 16-mm sequence for a play by Arthur Conan Doyle entitled, *The Mystery of Cader Ifan* (1967), starring Michael Latimer and Charlotte Rampling:[279]

> About two months ago I [Peter Sasdy] asked you [David Conroy] to make the necessary arrangements for a film cameraman for this forthcoming production. It will have unusually complicated film sequences and, being 'Wuthering Heights', will have to live up to feature film standards. […]
> Not long ago, for the play 'The Mystery of Cader Ifan', he shot 40 mins of synced 16-mm film in the Welsh mountains with which I was very satisfied. Since 'Wuthering Heights' will be shot under similar conditions on the Yorkshire moors and since John has proved himself to be extremely capable under such rugged conditions, I feel very strongly that he will be able to do it again and that I would like to have him.

At an early stage too (10th Aug), in order to organise the compulsory senior-management viewing of the 'play-backs' scheduled for the Autumn (27th Sept and 4th Oct), Conroy wrote a letter to the writer of the serial, prolific Irish playwright Hugh Leonard, in which he advised him of the dates, time (2.30 to 4.20 pm) and place (the office of BBC2 Head of Serials, Shaun Sutton, i.e., 'room 406 Threshold House' at Television Centre) for their meeting.[280] Leonard had received international acclaim (Prix Italia) for his Frank O'Connor-inspired teleplay, *Silent Song* (1967), and just finished adapting Charles Dickens' *Great Expectations* into a successful mini-series. He would be invited to the London-based rehearsals, then expected to be available for the duration of the studio-based multi-camera shooting, which was concomitant with the telerecording of the video footage on to film stock, and quickly followed by the editing and mixing of all selected footage, inclusive of the 16- and 35-mm filmed inserts (21st Aug to 21st Sept). In a letter signed by Carolyn Bill, another assistant to Sasdy (10th July), Leonard is not only asked to meet informally the director (20th or 21st July) to discuss the scripts with him before the departure of cast and crew for Yorkshire, but also to give 'a speech to cover telecine 3, episode 3, page 13'. In this speech, Heathcliff is to 'explain to young Hareton how to set a rabbit trap, perhaps ending "… now you try it for yourself"' so that the scene can be mixed 'to older Hareton actually doing it'.[281]

Meanwhile (12th July), Conroy was committed to the preparation of the shoot in Yorkshire, linking up with the North Region Programme Executive in Manchester to maximise the local news publicity deriving from filming some of *Wuthering Heights*' most emblematic sequences on location, without jeopardising safety, since the 'influx of crowds to watch the filming' on the moors would be synonymous with increased 'fire hazard', and could compromise the grouse season! Further, with the help of yet another of Sasdy's energetic assistants, Pennant Roberts who, after the director's reconnaissance of the ground and scouting for buildings in the Spring, was doing all the legwork in Yorkshire, Conroy officially secured Kildwick Hall for the exteriors of Thrushcross Grange. The producer also secured the 'derelict cottage' initially spotted by Sasdy for the production designer crew to revamp. This cottage was then used as a hunting lodge and was 'situated on the left-hand side on the road up from Ponden Hall', at a spot named Ponden Kirk.[282] This crumbling hunting lodge 'abut[ting] on to a vast area of grouse moor' and belonging to a Mrs Bannister, living at Carrhead Hall, near Keighley, would now 'impersonate' Wuthering Heights. The ever industrious Roberts took care of the first round of negotiations for the fees and insurance indemnities due to Mrs Bannister and all the landowners and leaseholders who would be potentially impacted by the filming at Ponden Kirk. He would also liaise with the Chief Constable of West Riding to ensure, well in advance, the security of the shoot as well as gain the trust of the local police.

What about Shibden Hall, its immediate garden and the remainder of its Park? The letter by Museums Director, Mr Innes, which pertains to the endnotes of Part II A, testifies to his effective transmitting of some fine local lore, since he is the one who tells Sasdy about the contrasting halls of Shibden and High Sunderland.[283] However, despite Innes' provisional agreement for a Sunday, early morning shoot (9th June), Shibden simply disappeared off the radar over the next few weeks: no filming seems to have taken place there after all!

The antagonistic houses of Thrushcross Grange and Wuthering Heights were not the only buildings to be represented. The Church of St Mary-le-Gill at Barnoldswick featured in the bright exterior sequence of Cathy's wedding to Edgar Linton in Episode 2 ('The First Revenge'), while the churchyard itself invited an original scene, absent from the novel. There Hindley Earnshaw/William Marlowe, maddened by the death of his young wife who has just given birth to their son Hareton, buries his head in the heap of fresh earth and wild flowers covering her grave. This scene served to stage a time ellipsis, as the next sequence, that of Cathy's wedding, starts off with a zoom-in on to Frances' now permanent stone grave. The bright day of Cathy's doomed alliance with the Lintons, in Leonard's serialised play, is thus commencing and finishing in the churchyard (not in the church), and is fully dramatising the Second MC (MC2). To allow the filming of those essential outdoor scenes, Conroy had gone as far as asking from Reverend Long the permission 'to dig a grave which is near the tombstone of John Dugdale, on the right-hand side of the small path between two Ewe trees',[284] and subscribed to his writer's artistic choice of, in a way, inviting the dead to Cathy's wedding. The shooting at Barnoldswick was finally optimised by the images, filmed in a close-up then from a distance in a variety of medium long shots, of Heathcliff/Ian McShane standing proudly by Cathy's grave, alone, in the immediate aftermath of her death, at the start of Episode 3 ('The Abduction').

The 1967 *Wuthering Heights* serial would be showcasing the work of a television trinity comprising drama producer, director and writer, in which Conroy and Sasdy were 'the figures of primary responsibility'.[285] While Conroy held the 'producer power' to finalise Sady's initiatives (for instance, the shift from 16- to 35-mm film for those horse-riding scenes involving dialogue),[286] Sasdy's role did not stop at scouting for locations, then directing the actors on location and in the studio, and supervising from afar the shooting-recording-editing process (film and VT). From pre-production to post-production, he had a 'hands-on' attitude towards the adaptative process. He was not in touch only with Leonard, but also very much the leader of a crew of specialised technicians, in particular, Peter Kindred, his production designer, whom he thanks heartily in a piece of correspondence dated 22nd Sept, which is part of Endnote 275. It was also Sasdy who came up with the numerous 'thank you' notes sent on his behalf by Ian Strachan (9th Aug), then suggested some 'Thoughts for Publicity' (22nd Sept again) to Conroy. Sasdy's advertising nous consisted in promoting *Wuthering Heights* through the achievements of his then enviable cast, Ian McShane who was to 'star in the Broadway production of *The Promise*', Angela Scoular (interpreting both Cathy and Catherine) who had two important films coming out, '*Great Catherine* (with Peter O'Toole and Jeanne Moreau) and *Round and Round the Mulberry Bush* (which [wa]s Clive Donner's new picture)', and last but not least William Marlowe, who had 'just received some excellent personal notices for his performance in the film *Robbery*', which could 'do the same trick for [him] that [sic] *Zulu* did for Michael Caine'. Conversely, 'some excellent photographs from location and studio' pertaining to *Wuthering Heights*, argued Sasdy, could be put to good use to actualise their busy agendas, and allow the *Classic Serial* to enter in a dialogue with the other two leading audio-visual artforms of the time, theatre and cinema.

The 1967 *Classic Serial* reveals the BBC's growing confidence in the fulfilment of its technological, dramatical and educational ambitions. There is this bold combination of 16-mm film to 'enable dialogue sequences with horses moving on rough moorland to be shot, using hand held cameras' (memo of 18th May) with the traditional but dearer 35-mm monochrome film dedicated, for instance, to the snow effect sequence that is mentioned in a letter to Wyverns Productions dated 31st Aug. The snow sequence is itself to be 'superimposed over the 16 mm film of Cathy's ghost', on 15th Sept, at an early post-production stage.[287] Further, in the 18th May memo when Conroy, on behalf of Sasdy, requests 'a change from shooting the location sequence for *Wuthering Heights* from 35mm to 16mm' (18th May), the producer adequately mentions 'the rising cost of this programme due to the expansion of its film requirements' in order to finish convincing, we can safely assume, Head of Serials, Shaun Sutton, of the unusual necessity for 16-mm footage within a *Classic Serial*. 'Shooting on 16-mm was acceptable for newsgathering and documentary but not for prime-time drama'[288] until the ground-breaking filmed drama, *Up the Junction* (Garnett and Loach, 1965), pushed the formal and thematical limits of *The Wednesday Play* by resorting, essentially, to 16-mm location filming and paved the way for another masterpiece of heartfelt social realism, *Cathy Come Home* (Garnett and

Loach, 1966). In 1967, the two BBC Audience Research Reports on the *Wuthering Heights Classic Serial* were not so much characterised by a sentiment of uneasiness about a change in the appearance of the image within the space of a couple of sequences involving studio and location filming. The coexistence of cinematic and televisual images with a different grain and background texture was not so much what was perceived as a stylistic hurdle as the mixing of the different soundtracks and their synchronisation with the image, where the wind effects were concerned. Accordingly, a viewer cheekily commented:

> There is not a howling wind in Yorkshire throughout the year and if it does blow, the trees move.
>
> (Week 47 AR)

What the two BBC Audience Research Reports render very much palpable too is that despite a strong awareness to the feigned Yorkshire accents of the cast (Week 44 AR), and to the compression of the story into four episodes (both ARs) and its associated 'breaks in the passage of time' (Week 44 AR), 'the balance of opinion was […] very much in favour of the whole serial' (Week 47 AR). The conventions of seamless visual transition and imperceptible mixing of soundtracks, stamping Classical Hollywood cinema and 'the film version with Laurence Olivier as Heathcliff' (Week 47 AR), were not really what mattered; the attention of the viewers was centred on the four televisual performances themselves, and they had been generally enjoyed with a Reaction Index of 70 for the last episode. The adjustment to the televisual formula of the *Classic Serial* had happened over more than a decade with, most notably, the frequent adaptations of Dickens' novels.[289] The simple, flatly chronological serial format of *Wuthering Heights* chosen by Hugh Leonard in 1967 required just a little intellectual exertion to follow the transgenerational, duplicated plot (Week 44 AR) in which Angela Scoular symbolically interprets the dark- and fair-haired Cathy/Catherine, as Juliette Binoche would twenty-five years later. What Hugh Leonard and Peter Sasdy's *Wuthering Heights* demanded from the television audience was a good measure of Coleridgean 'suspension of disbelief'[290] since the *Classic Serial* was still defining its relationship to continuity editing and fine-tuning its own stylistic norms: those norms could be neither Classical nor 'Free Cinematic'. The viewers, of necessity, were impelled to develop their viewing and critical skills while contributing sometimes, through their feed-backs collected for the BBC Audience Reports, in building a comprehensive television aesthetics 'adducing reference points from contemporary theatre and cinema'.[291] This would give rise, in due time, to a more sophisticated serialisation of *Wuthering Heights*.

Both conceptually and technically, the 1967 production was innovative with location filming on film and studio multi-camera shooting on tape, which were carefully planned and executed, and entailed working with two specialised editors, the film editor Bob Rymer (who is the only one mentioned in the *Wuthering Heights* credits) and the VT editor Geoff Hicks from Wyvern Productions. The Ampex VT technology, now taken to its full capabilities with VT editing, would progressively help in moving away from the teleplay on to a credible serial format. At the request of Sasdy, who had 'always been very satisfied with his work' in the past, the same video editor, Geoff Hicks, was retained for all four episodes, 'the reason for this request [being] that the unity of the editing style of this show [was] so important and because it [was] so complex'. This memo, signed by Carolyn Bill and dated 5th Sept, 1967, mentions the editing dates of 18th, 19th, 20th and 21st Sept given by Planning, which means there was a mere five- to six-week interval from post-production editing to the transmission of the first episode on BBC2 (28th Oct). Through this *Classic Serial* in which the emblematic outdoor scenes were shot on 16-mm film, and the more theatrical scenes were proudly shot electronically in studio, the BBC was expressing its endorsement of a televisual aesthetics of contrast in the image texture, which could serve the *mise en scène* of an ambience of heightened tension at some epiphanic moments, such as the immediate aftermath of Heathcliff's death. This intense dramatic moment translated into a sequence of sheer theatricality with the broad sweep of the camera, which first closed up on to the bloody wrist of Heathcliff, lying dead in the haunted bedroom, then moved on to a prolonged shot of Cathy's portrait, nestling high up in the living-room dresser, and concluded with a wide-angled, open perspective on to the entire space of an emptied studio stage. It is only after this necessary studio finale that the action could return to the wind-swept moors where a now freed (and probably married) young couple, Catherine and Hareton, is seen walking back towards their happy dwelling, accompanied by Joseph, Ellen and the little shepherd boy.

The hint of condescension one can discern in David Conroy's response to some of the criticisms of the viewers, as he champions his team's dramatic and aesthetical choices, not only signals the protective attitude of the BBC2 *Classic Serials'* producer towards the hard work accomplished, but also seems to be a side effect of the BBC's cultural prerogatives which, just then, a BBC executive would have regarded as unquestionable. Conroy's last lines, in his reply to well-intentioned reader of classics, Miss Vanessa Clark, are quite revealing in this respect. She did not believe in the team's *mise en scène* of Heathcliff's intense grief, in the orchard, at the death of Cathy:[292]

[…] As regards closing his eyes this was in fact as a result of the exhaustion he felt from the height of his passion. I am sorry if this was not apparent to you.

I hope that you will look at future dramatisations of classic novels with a slightly new eye now that you realise some of the problems involved in presenting them on television.

From the perspective of a television historian, the appeal of this four-episode serial is not lessened in any way by the lack of psychological depth in the Lintons or the absence of a coloured rendition. Back in 1967, the spectacular improvement in the dramatisation of the Gothic classic, *Wuthering Heights*, on television was not only complemented by the strong impersonation of Heathcliff by Ian McShane – the wild Harry of *The Wild and the Willing* (1962) – but also by the arresting presence of a young Ellen portrayed by Anne Stallybrass, the endearing actress who, four years later, would be starring in *The Onedin Line* (1971-1980). Stallybrass was rooting in the heartfelt emotion of a privileged insider, Ellen/Nellie, the difficult *mise en scène* of Heathcliff's *sublime* moments of grief and torment, as well as superseding by her staying power and quality of empathy the Lockwood of Jeremy Longhurst whose ten-minute appearance in Episode 4, and recapitulative voice-overs at the beginning of Episodes 2 and 3, would not be quite sufficient to transform him into a credible intra-diegetic observer. As for the *beautiful* and its actual rendition on the small screen, it was made particularly vivid, thanks to some very nice touches added to the original screenplay. Inspired by the shoot on location, Leonard and Sasdy had been imagining Cathy's close interaction with toddler Hareton, at St Mary-le-Gill churchyard, and Ellen's caring impulse for the frightened shepherd boy, whom she would gently invite to share the comfort of a now pacified Wuthering Heights, at the end of the play. If these considerable translative achievements cannot hide the lack in representation of the Gothic Figure of *Mise en Abîme*, they contributed to the make-up of an expandable template that would be later re-shuffled into a five-episode serial, and re-imagined in colour in order to adapt to the regular capabilities of television broadcasting and reception in the late 1970s.

The translation of the natural outdoors on film was not always perceived by the contemporary viewers in the Autumn of 1967. This lends a particular resonance to the testimony given, thirty-four years later (November 2011), by director Andrea Arnold to *The Telegraph*'s film critic Benjamin Secher. Talking about the filming on location, Arnold was deploring the more than partial loss, on the cinema screen, of the muddiness and exacting wilderness she had felt in all the fibres of her body, on the North York moors.[293] Still, through the fervent rapport of her Heathcliff with the raw beauty of the moorland, Arnold's controversial *Wuthering Heights* comes across as one of the most powerful renditions of the First MC (MC1) on 16-mm film. Similarly, the first *Classic Serial* remains, with its integration of the contrasting landscapes and houses into the screenplay and its careful *mise en scène* (both reflecting on the BBC Audience Research Reports) a harbinger of the actualisation of MC1 on television. Further, the serialised *Wuthering Heights* of 1967 with its four episodes ('An End to Childhood', 'The First Revenge', 'The Abduction' and 'The Last Revenge') started to add some temporal texture to the rendition of Emily Brontë's novel on screen as it aimed for a more exhaustive translation than the one that had been envisaged before in the Kneale-Cartier teleplays – or even the 1939 feature film. This constructive attitude towards the original text and its unconscious would benefit Sasdy and Conroy's next *Classic Serial*, *The Tenant of Wildfell Hall* (1968), beautifully adapted from Anne Brontë's epistolary novel by famous playwright Christopher Fry. There would be, for *Wuthering Heights* on the small screen, no further attempt to either eclipse the early generation of children or ignore the second generation of lovers. For the vitality of MC3 on screen, it was already felt (and implicitly demonstrated) that the lovers' transgenerational journey had to somehow influence, or even motivate, Ellen's and/or Lockwood's self-revelatory journeys. Ellen and Lockwood were beginning to be seen as equally important characters, in the sense that she could now shed her slough of self-righteous commentator, as Lockwood could abandon his slough of impassive onlooker. All this would be contributing to the simultaneous apprehension/revelation of the imaginative sphere surrounding the Third MC (MC3).

While the other programmes on BBC1 and ITV had considerably more success, and the size of the audience for the serial's fourth episode ('The Last Revenge', Week 47) was halved at 2.7% of viewers compared to 6.2% for the first episode ('An End to Childhood', Week 44), all the efforts of the BBC2 *Classic Serials* team were in the end rewarded by some excellent feed-back from the sampled viewers of the BBC Audience Research Reports:

> [...] they had found 'The Last Revenge' a particularly exciting episode, full of action and drama, and rather easier to follow than some of the previous ones; 'it was more concise and took place in a shorter period of time'. They had evidently been 'moved' by the death of Linton and 'gripped' by the 'spine-chilling' appearances of Cathy's ghost: 'This fantastic and horrible story was superbly well done'.
>
> (Week 47 AR)

*   *   *

Archival References

1978 Programme:
RCONT21 – David Snodin Copyright Contributor File
BBCWAC T48/374/2 Hugh Leonard (TV Script Unit File)
BBC Audience Research Report on 'Wuthering Heights', tx. Sunday, 24th Sept, 1978 (Week 39) 8.05-9.00 pm – R9/7/156
BBC Audience Research Report on 'Wuthering Heights', tx. Sunday, 24th Sept to 22nd Oct, 1978 (Week 43) – 8.10 pm approx. – R9/7/155

In its second and fourth episodes, the serial composed by Hugh Leonard and *mise en scène* by Peter Sacsy succeeded in conjuring up the Gothic on to the small screen. As I have previously pointed out, the Second MC (MC2; Revelatory Dreams and Imagination-Charged Visions; Connection between the Living and the Dead) was made very much alive in the outdoor sequence of Cathy's wedding at Barnoldswick. Further, as relayed by the last excerpt from the BBC Audience Report (1967, Week 47, Episode 4), the Gothic was the main characteristic of the indoor nightmare sequence where a well-mannered Lockwood, interpreted by Jeremy Longhurst, quickly decamps from the Heights, not to be seen or heard again, after his ordeal of seeing Cathy's ghost in a 16-mm film insert that contained snow effect shot on 35-mm, and required full suspension of disbelief! The television viewers would have to wait another decade (1978) for Lockwood's nightmare to feature in Episode 1, and for Lockwood himself to become an active intra-diegetic observer facilitating a more even re-surfacing of the dynamic structures of Emily Brontë's novel across a whole *Classic Serial*.

Richard Kay incarnated the 1978 Lockwood. Through his participation in the second episode of Christopher Fry's *The Brontës of Haworth* (1973), he would have evoked the Imaginary world of the Brontë family in the minds of the most alert viewers. His multi-layered Lockwood marked the bold, creative début of a young television writer, David Snodin. As shown by the 'Agreement relating to the commissioning of specially written television dramatisations' (which was destined to Snodin, the newly contracted writer) and as confirmed by the 'Existing Material Brief' signed on 21st Feb, 1978 by Head of Drama Series and Serials, Jonathan Powell, Snodin was entrusted to recast Leonard's initial template from 1967 into a script that would increase the original screen time, 180 minutes in total, by more than a third with five episodes of fifty-two minutes each. The educational remit of the newer BBC2 *Classic Serials* was still much the same as in 1967: promoting education through the re-discovery of literary classics exhaustively adapted on to the screen for BBC Licence Fee holders.

The translation on to the screen of Emily Brontë's *Wuthering Heights* (and its unconscious) would clearly be favoured again over the more superficial readings of the Wyler movie, and adaptation would be put right at the centre of the scenarist's and director's preoccupations. With the help of accomplished director Peter (Charles) Hammond, Snodin re-dynamised the old master's ideas for those audiences who were ready to engage with the vivid style of a colour *Wuthering Heights* on television, and had delivered themselves from the emotional hold of the black-and-white Wyler movie. They would be facing the challenge of a Lockwood promoted to the position of intra-diegetic observer and actualiser of the Gothic Figure (*Mise en Abîme*). The dynamic structuring of the novel, inclusive of the Bataillan Themes, would effectively be brought back to the surface after Snodin's astute re-shuffling and augmentation of Leonard's 1967 script.

Unsurprisingly, some of the viewers sampled for the 1978 BBC Audience Research Reports considered Episode 1 'too miserable, too depressing and too gruesome' (Week 39 AR) and would be reticent to make the mental adjustments necessary to digest the full format of an unabridged adaptation. They would not watch at close range the re-emergence of the MCs and BTs in the remaining episodes, and would miss out on the enjoyable presence, in Episodes 4 and 5, of a vibrant Catherine who was neither conceived nor interpreted as a pale replica of Cathy. The halving of some already modest audience numbers from Episode 1 (5.8%) to Episode 5 (3%) would be counterbalanced by a series of steady Reaction Indexes (66, 65, 66, 65 and 62):

The production as a whole fared fairly well, without perhaps being considered outstanding. However, many viewers commented on how well the atmosphere of the moors had been captured and, indeed, the mood of the book.

(Week 39 AR)

This would not deter the BBC from hiring Snodin again and again. Trusted with his constructive attitude towards recasting Hugh Leonard's *Wuthering Heights*, he would rapidly progress, notably as writer and script editor for the prestigious series dedicated to Shakespeare's plays (1980-1984), then as producer of high-profile screenplays, both adapted and original, *The Men's Room* (1991), *Great Expectations* (1999) and *Tess of the D'Urbervilles* (2008), amongst others. Back in 1978, nonetheless, the bright re-vision of Hugh Leonard's script by David Snodin was not the only ingredient for the spark in the new production of *Wuthering Heights*; the popular television swashbuckler of the 1950s, Peter Hammond, contributed to it through his well-rounded *mise en scène*.[294]

Hammond had already re-invented himself as an acclaimed television director rewarded, by the middle of the 1960s, by the 'Guild of Television Producers and Directors' and was on a par with Peter Sasdy for opening up the space of the television studio to suspense and mystery.[295] At quite an early stage, he handled with great agility the *mise en scène* of his episodes of *Out of This World* (1962) and *The Avengers* (1961-1964) throughout which one can detect an evolution of the visual grammar of the television medium towards more complexity. At that stage, Hammond had also tried his hand at the *mise en scène*, for the BBC, of a well-loved classic, *The Count of Monte Cristo* (1964), and directed its first episode. Hammond was so intent on exploring the aesthetical and dramatic potential of VT recording that he then declined the offer of Brian Clemens, the newly appointed producer on *The Avengers*, for the next series that was to be shot on film. Hammond had opted instead to make the mini-serial, *Contract To Kill* (1965), a thriller staging a Nazi hunter, that he would complement with a swashbuckling mini-series entitled *Hereward The Wake* (1965). Following his inclination for the liveness and crispness of the teleplay, he would receive special applause, on the same year, for the spine-chilling drama *Ambrose* (1965), a forerunner of Robert Altman's *Images* (1972), which he directs for Leonard White's *Armchair Mystery Theatre*. He was to carry on with some serialised productions that he would direct entirely on his own such as *The Three Musketeers* (1966-1967), *Our Mutual Friend* (1976) or *even* a remarkable Gothic thriller *The Dark Angel* (1989) starring Peter O'Toole and largely inspired by Sheridan Le Fanu's *Uncle Silas*.

Finally, Hammond's portrait would not be complete without mentioning his directing of some of the most fondly remembered episodes in the *Sherlock Holmes* series – *The Sign of Four* (1987), for instance – as well as of the first two episodes of a Fantastic tale entitled *King of the Castle* (1977). There, and this would be a good exercise for *Wuthering Heights*, Hammond's *mise en scène* credibly superimposed the stark council estate reality of little Roland's bullying – and his deep feeling of estrangement – on to a parallel world of excitement and adventures, in which the boy fully takes control of his life and surroundings.

For the *mise en scène* of his outstanding *Classic Serial* (1978), Hammond would be using the outdoors as background for an action synchronised with the pace of racing horses and fast carriages, or unfolding much more slowly and strenuously owing to immutable gates, portals and stone walls. He would also be transforming the outdoors into a finely woven, multicoloured tapestry where the First MC (MC1) can firmly take hold of the small screen, as its colours are enhanced by the contrasting soundtracks of 'Cry of the Ghost' and 'Sun as Fire, Light, Gold'. The faraway moors and changing skies appear to be animated by a kaleidoscopic camera lens, playing with some carefully foregrounded motifs of heathen life. By contrast, in Hammond's interior tableaux, Wuthering Heights remains in a vestigial penumbra accentuated by Carl Davis' darkest themes – 'Notes for the Satanic Subconscious', 'The Dark at the Top of the Stairs' and 'Lockwood's Nightmare' – while the grand and cushy interior of Thrushcross Grange is inundated by some heavily diffracted streams of light matching the destructured score of 'Intimations of Madness' and foreboding Cathy's mental break-down after the devastating row between her husband, Edgar, and Heathcliff.[296] This excess of strong iridescent light changes the plants and flowers adorning the great drawing-room into a drapery of lush and stifling verdure that will soon shroud entirely Cathy's body as she lies there, in broad daylight, in a pose strongly reminiscent of the drowned Ophelia of John Everett Millais' celebrated painting (1852).[297]

Hammond's 'trademark style', an expression chosen by television historian Dick Fiddy in his elegy to the late director,[298] consisted in composing each scene like a tableau vivant with, from foreground to background, shaded layers of dramatic and visual interest that would be progressively revealed. This was allied to his propensity, when staging epiphanic moments in studio, to combine the image seen through the lens of the VT camera with its reflected double captured in a mirror or glass pane. In his television-designed *Wuthering Heights*, these auteurist traits set off the scenes of *sublime* intensity when Cathy, in 'her feverish bewilderment', fails to recognise herself in her chamber's mirror[299] or when Heathcliff, in the immediate aftermath of Edgar Linton's death, takes possession of Thrushcross Grange library and confides to Nellie, who only dares to look at him indirectly through the mirror hanging above the fireplace, how he found a little respite from his tantalising quest by opening the lid of Cathy's coffin, and dreaming he was lying by her side.[300] In those difficult passages closely translated from the source text, Hammond's sophisticated camerawork and *mise en scène* reinforce the appeal for long, detailed sequences and place the picture itself, in a Hepworthian (or Stannardian) 'pre-classical Hollywood' fashion, at the 'centre of the pleasurable perception'.[301] His idiosyncratic visual style would allow the unequivocal re-surfacing of the Second and Third BTs (unbearable tension between Hyper-Morality and Morality; Physical Death and Re-Generation). This would be a first for a television adaptation of *Wuthering Heights*. As underlined in the

BBC Audience Research Report assessing the impact of the first episode (Week 39) – there would be another report put together four weeks later, after the fifth episode had been broadcasted (Week 43) – this new *Wuthering Heights* directed by Hammond and scripted by Snodin-Leonard felt quite foreign to a substantial number of spectators in the viewing panel, despite its compelling aesthetics linking up with the MCs and BTs:

> In spite of the fact that a majority of the sample audience said they had read *Wuthering Heights*, one of the main complaints was that this version was hard to follow. Viewers apparently had some difficulty in sorting out the characters and their relationships. Over a quarter found the order of the narrative confusing and there were further criticisms that the whole episode was disjointed. Perhaps as a result of this confusion a majority of viewers failed to feel really involved with the people in the story and had little sympathy with their situation. It was clear that the Oliv[i]er film version was for some the definite adaptation, and tonight's episode was compared unfavourably with it, particularly in terms of overall atmosphere. […]
>
> (Week 39 AR)

In the opening episode, David Snodin's screenplay was replicating, without leaving out many a page, the sequence of events as presented in the first seven chapters of Emily Brontë's novel: from Lockwood's introduction to the inhabitants of the Heights – and his nightmarish vision of the ghost-child Cathy – in the early Winter of 1801, to the flashbacks originating in Nellie (and re-imagined by Lockwood and the extra-diegetic observers), which relate the arrival of the abandoned child Heathcliff in the old Earnshaw family in the late Summer of 1771, as well as Heathcliff's confinement in the garret after his thrashing by Hindley on Christmas day 1777. The difficulty experienced by some viewers in deciphering Episode 1, and their subsequent rejection of the whole serial, could quickly be interpreted as the side effect of a succession of cinematic flashbacks where some convoluted events are recounted visually without the support of a reassuring voice-over. However, this difficulty could also be ascribed to those viewers' own denial of the text and subtext of the novel, since the opening episode's emphasis on the Second MC (MC2; Dream Visions of Cathy as a ghost-child, of Heathcliff as a ghoul; Correspondence between Childhood and the Sacred) and its insistence on the downside of the First BT (BT1) – with, on the one hand, a Lockwood seen wading through a Slough of Despair and grappling with his inmost fears and, on the other, a little Heathcliff seen enjoying wildly the 'Ephemeral Sovereignty of Childhood' despite his trashings and beatings – are both genuine and adequate. This emphasis on MC2 and insistence on BT1 in Episode 1 must have contributed to the feeling of confusion and disorientation of some of the most vocal viewers in the sample audience.

Since the audience ratings were halved from Week 39 to Week 43, it means that only half of the sampled spectatorship was able to watch the entirety of Lockwood's Self-Revelatory Romantic Journey. Those who did see the five episodes of the serial (even though they were considerably less vocal than those who did not!) could not have been insensitive to the progressive unveiling of the much brighter sides of the Third Mythical Component (MC3), culminating in a very happy ending for the second generation of lovers. They would also have been able to experience a new sense of empathy for the darker heroes, with the resurgence of all three BTs, as the serial was drawing to a close. In Episodes 4 and 5, the liberating aspects of MC3 were showcased by the strong performance of a particularly well-cast actress, Cathryn Harrison, who interprets young Catherine following her much commented appearance in Louis Malle's *Black Moon* (1975) as the adventurous teenager Lily. Supported in her impersonation of Catherine Earnshaw by Yorkshire-born actor-producer David Nicholas Wilkinson (young Hareton) and previously mentioned Richard Kay (Lockwood) who would be last seen in an episode of Ian McShane's *Lovejoy* (1986), she illuminated the small screen by effortlessly finding her place, then standing her ground, in the stark feudal dominion of a Wuthering Heights controlled by Heathcliff. In the opening flashback that lasts nearly twenty minutes, at the beginning of Episode 1, Kay has just had enough time to introduce Harrison to the viewers before his blood-chilling encounter with the ghost of her mother. His re-appearance in Episode 5 corresponds to a re-play of his ghastly ordeal of Episode 1, which is however much shortened with a re-edited sequence lasting 4.30 minutes only! This re-play, which is also re-inserted chronologically in the serial's diegesis, is expressly designed to focus on the re-actions of the inhabitants of the Heights, mainly Heathcliff's but also Catherine's and Hareton's.

Therefore, the successful rendition of the three-dimensional Gothic architecture of *Mise en Abîme* is achieved by the repeated, yet slightly altered, *mise en scène* of Lockwood/Kay in a crucial sequence, common to Episode 1 and Episode 5, his arrival at the Heights. In Episode 1, Wuthering Heights is seen through his eyes, the eyes of a detached intra-diegetic outsider who will soon strike an initiatory attachment with the daughter of the ghost, Catherine/Harrison, and start off his Self-Revelatory Romantic Journey. In Episode 5, the perspective is inverted. Lockwood, with whom the viewers have now identified, has undergone a transformation and is seen through the eyes of the dwellers of the Heights who also have, all three of them, nearly completed their Initiatory Love Journey. In 1991, Anne Devlin would borrow David Snodin's innovatively faithful cinematic translation of the architecture of *Mise en Abîme* for her *Wuthering Heights* script.[302] In 1998, it would be Neil McKay's turn for David Skynner's television version. In Snodin and Hammond's secretly celebrated version, the re-play of Kay's introduction to the Heights (and interaction with the ghost-child) is soon followed by an elliptic sweep of the camera over moors and skies. Then and then only, his second actual exploration of the Heights, as opposed to his dreamlike exploration of the Heights, is being staged extensively.

In early January 1802, two months after his 'sinking up to the neck in snow' on his way back from the Heights, Nellie would entrust him with 'a little note' destined to Catherine.[303] This second visit, being flirty and light-hearted, is rather pleasant for

Lockwood/Kay. His short-lived certitude that he is a suitor superior to any Catherine could meet in such a remote location provides comic relief to the extra-diegetic observers, who have already discerned (even without the warped recountings of the housekeeper Zillah) a change in Catherine's attitude towards Hareton, her prince in disguise, and a change in her perception of herself. After the deaths of her father and young husband, Linton, Catherine becomes a narrative force that accelerates the transformation undergone by Heathcliff and Hareton, and facilitates the narrative development of those intra-diegetic observers who are amenable to changes. Therefore, she not only intervenes in the mythical layer of Cathy and Heathcliff's tale, but also functions as a catalyst for Lockwood's romantic story, which is situated at the topmost level of the nested frame of narration. Materialising both as Hareton's and Lockwood's love interest, she gives the extra-diegetic observers some rewarding instants of Comedy and Romance to enjoy.

* * *

The flair that Peter Hammond demonstrated in actualising David Snodin's innovative televisual scripting of the novel's *Mise en Abîme* was associated with a breath-taking interpretation of Catherine by Cathryn Harrison. The effectiveness of Hammond's *mise en scène* in Episode 5 thus depended on the building up of a propitious creative tension between actors, and between actors and director, although the chemistry seemed to have existed from the very start between Harrison, Hutchison and Wilkinson. All this resulted in the coming alive of the First BT (the Present Moment and the Sacred) and in the facilitation of the viewers' eventual adherence to the Second (Empathy with Heathcliff's Plight) and Third (undergoing, together with Lockwood, a Moral Transformation that changes one's Life) BTs. In particular, it guaranteed the pervasiveness of the 'light weight' of the Present Moment in the amusing sequence of Hareton's compulsive pipe smoking, which features in Hugh Leonard's 1967 screenplay. This sequence announces Hareton/Wilkinson's subsequent reconciliation with his cousin Catherine/Harrison over the accepted book, which translates very closely the novel's narrative sequence of events.[304] Hammond's collaborative work, through its reliance on the Present Moment, bears a lot of similarities with Jacques Rivette's epiphanic *mise en scène* of *Hurlevent* (1985) and David Skynner's naturalistic vision of love in his staging of the second generation of lovers (1998).

In the 1978 *Classic Serial, Wuthering Heights*, the balance of characters that Snodin achieves in his expansion of Leonard's script was not unsettled by a weak casting of Edgar or Isabella Linton who are impersonated by two charismatic actors, David Robb and Caroline Langrishe. The former would have been immediately recognisable, at the time of broadcast, as the Germanicus of *I, Claudius* (1976) while the latter would have been unmissable as both the Kitty of the *Classic Serial, Anna Karenina* (1977), and the Cosette of the television version of *Les Misérables* (1978). If the balance of characters was unsettled, it was not at the level of the heroes or anti-heroes, but at the level of the privileged insider, Nellie. Without her mitigating influence as an intra-diegetic observer who, like the Ellen hinted at by Anne Stallybrass in 1967, would be willing to follow her own narrative arc, the re-surfacing of the novel's unconscious (in the episodes staging Cathy and Heathcliff's tormented passion) comes across as too brutal and unnatural for some sample viewers who 'comment[ed] on the amount of shouting that went on' (Week 39 AR). Contrary to David Conroy and Peter Sasdy who had opted for a genuinely Brontë-esque Ellen/Nellie belonging to Hindley's age group, Jonathan Powell's *Classic Serial* somehow read her as an unadventurous woman, and cast Pat Heywood in her role. Heywood's part as Juliet's nurse in Franco Zeffirelli's *Romeo and Juliet* (1968) was clinging to her interpretation of Nellie like a second skin, and transformed her into an irreproachable stronghold immune to love. She would be at the opposite end of the dramatic spectre of the vibrant Ellen incarnated by Stallybrass in the 1967 *Classic Serial*, as well as lacking an agenda of her own. This characterisation proved to be a translative flaw as regards the dynamic structures of story-retelling since, unlike Stallybrass with Angela Scoular, Heywood would be unable to do justice to Kay Adshead's sensitive and clever interpretation of Cathy; in particular, she would be unable to arise the viewers' feelings of empathy for the distressed heroine, at the time of her second illness. While bearing the mark of a life-threatening pregnancy in the intended *mise en scène*, that illness would often only be coming across as mental and self-inflicted on screen.[305]

Owing to this too common revisitation of the character of Nellie, which does not take into consideration the fact that she also belongs to the abysmal narratorial structure and oscillates between Rejection **and Identification** in her embedded narratives, Cathy's 'hysterical' break-down would be associated, for some sample viewers, with Hammond's unrestrained *mise en scène*, and Adshead's over-interpretation. This is recorded, once more, in the first BBC Audience Research Report (Week 39):

> There was a slight feeling that the acting tended towards the 'hysterical' and a few reporting viewers commented on the amount of shouting that went on.

Nonetheless, neither the *mise en scène* per se nor the interpretation of actress (and future playwright) Adshead was at fault. With Mike Leigh's *Kiss of Death* (1977), her perfectly controlled acting had already been put to the test of smouldering scenes. The BBC Written Archives reveal nothing about the balance of power, on *Wuthering Heights*, between Powell and Hammond. The latter was revered by Bob Fuest, the director of the 1970 *Wuthering Heights*, as a television practitioner who had perfected a style of *mise en scène* that actualised the Present Moment and involved a pictorial, and even sensorial attention to detail, in the composition of all sequences.[306] The adequacy of the 1978 *Classic Serial* in terms of overall 'atmosphere' and 'period atmosphere' was somehow acknowledged in both BBC Audience Research Reports, which is significant of the accession of the 1970s audiences to the aesthetics of the period drama:

> [...] many viewers commented on how well the atmosphere of the moors had been captured and, indeed, the mood of the book.
>
> (Week 39)

> [...] the sets, outside filming and costumes were all praised in some quarters and it was thought that the production had succeeded in capturing a very realistic period atmosphere.
>
> (Week 43)

B

The 'Period' Dramas and the 'Anti-Period' Dramas: Reflections on Cultural 'Accuracy' and Cultural 'Displacement' as Inspired by Steiner's Hermeneutic of *'Compensation'*

Chapter 24

The British Period Dramas - From One Generation to the Next, from Hollywood to ITV: *Accuracy* and Compensation

Fuest-Tilley (1970, AIP, 104 min)
Kosminsky-Devlin (1992, Paramount Pictures, 105 min)
Skynner-McKay (1998, ITV1, 112 min)
Bowker-Giedroyc (2009, ITV1, 137 min)

By 1969, the indigenous energy of the British New Wave project (*Heathcliff* or *Love for Life*) had already rippled through the *Wuthering Heights* made for the 1967 BBC2 *Classic Serial*. It had also increased the relevance of a modern period drama promoting Keighley and its Georgian industrial history to the category of thematic component, and the inhospitable Yorkshire moors to the status of mythical component. By weaving contrasting settings and characters into a socio-historical canvas where an old feudal micro-society was superseded by a system, much more subtly ruthless, born of colonisation and industrialisation, the scenarist of the 1970 *Wuthering Heights* was, concretely, providing the film narrative with an array of protagonists, ranging from masters to servants, and evolving according to parallel agendas and incompatible destinies. The narrative arc of the 1970 British feature film, strengthened with its historical reflection on class, social ascension and patriarchy, came to complement the serialised format of the 1967 British television production, which could not boast of an ideological subtext, but had allowed for an extensive and very rich translation of Emily Brontë's novel on to the small screen.

Materialising another translative strategy altogether, the classic Wyler picture had effected a cultural displacement by staging a fatal love affair on a romanticised backdrop of wild heath and glamorous outfits suiting Merle Oberon to perfection, and by inventing a rather stately Wuthering Heights complete with neatly kept stables where the degraded Heathcliff/Olivier suffers from the injustice of being seen as a beggar. This set-up conditions Cathy/Oberon's mercenary behaviour and gives rise to her brutal admonitions, until her avowal to Ellen of accepting the proposal of the smug Edgar/Niven, overheard by Heathcliff/Olivier, finally propels the latter outside of the Heights for his transformation into a rich man. British graphic designer Patrick Tilley, who could not care less for the Olivier version, would become the scenarist of the 1970 *Wuthering Heights*. Having been noticed as production illustrator and script doctor on *Oh! What a Lovely War* (1968), American International Pictures (AIP), an American production company of Z movies hungry for respectability, chose Tilley to adapt the English literary classic by Emily Brontë: it would be honed in Yorkshire – three weeks on location! – and completed in the Pinewood Studios, near London, by a British director, a British crew and a British cast. For the writing of his script, Tilley intended to get back to the basics of the novel through the combined readings of David Cecil and Somerset Maugham.[307] Nonetheless, he was to find his most original creative strand in an unusual secondary source, *L'Eau des collines* (1963) – a two-part novel by cinema lover, acclaimed cinéaste and writer Marcel Pagnol. *L'Eau des collines* was first imagined and constructed as a movie, *Manon des Sources* (1952), before being completed with its prequel, *Jean de Florette*, as a solid literary text, a decade later:

> I read the novel five times, David Cecil's study of the novel and some other related material – including a timing of the events by David Daiches. In his foreword of the 1965 Penguin English Library Edition, he establishes the ages of the principal characters. Discovering that Heathcliff was 17 and Cathy 16, and that it was a Georgian (and not a mid-Victorian) period novel made a great difference. It helped me make sense of what turned out to be a deeply complex story. I likened it to peeling back the layers of an onion. […]
>
> My life was transformed by the works of Marcel Pagnol, and the scenario of *Wuthering Heights* (also entitled *The House on the Moor*) was probably influenced by my reading of *Jean de Florette* and *Manon des Sources*.

Tilley was drawn to the transgenerational family saga recounted by Nellie Dean to Lockwood, then by Lockwood to the readers. This family saga, starting for good half-way through Vol. I as Hindley's son, Hareton, is born,[308] had nicely been adapted by Tilley in the comfort of his home until the entire pre-production team (inclusive of Deke Heyward's thirteen-year-old daughter!) dissuaded him from relying on Vol. II, which sees the unfolding of Hareton's story with his two cousins, Catherine and Linton. Since the AIP producers, Arkoff and Nicholson, only coveted the prestige that the previous (truncated) version had brought to Goldwyn, it made sense for their team to develop fewer characters better (the ones carrying the dramatic weight of Vol. I) so as to give them more room into a movie that does not exceed 104 minutes. Tilley would drop Vol. II with the exception of its first two chapters revolving around Cathy's death, as witnessed by Ellen.

Robert (Bob) Fuest, the appointed director, and a painter trained at the Wimbledon School of Art, had been drawn to the *sublime* dimension of the first half of the Initiatory Love Journey (from Eros to Thanatos), and it was not too difficult for him to convince Tilley, a graphic designer at heart, of the tangible advantages of a condensed version of the novel as well as reassure him about its visual impact on the imagination of the viewers. Further, the Buñuel picture, which they had viewed together shortly after AIP hired them, had made a strong and lasting impression on them both; they felt united by the conviction that they should try and emulate its aesthetics.

Considering that Lockwood's love interest, the second Cathy/Catherine, had vanished at the same time as Vol. II, Tilley rightly deduced that the love journey had been deviated from its romantic trajectory. In Tilley's new scenario, Lockwood became redundant both as under-cover suitor and informed spectator. To *compensate* for Lockwood's absence, Tilley intuitively upgraded the character of Ellen so that she could fulfil his mission of intra-diegetic outsider, whose 'trespassing and learning' not only serves the plot but also encourages the gaze of the extra-diegetic viewers (or 'spectators') – as in the *Tristan and Iseult* stories. This meant, for the Tilley-Fuest tandem, working out a 'translative strategy of **compensation**' where Ellen is given the screen presence of an Individual character no longer limiting oneself to playing the part of an Adjuvant character – such as the 'helper' or the 'dispatcher' identified by Propp in his *Morphology of the Folktale* (1928). The privileged relationship between Ellen and Hindley, as well as Ellen's privileged position inside the Earnshaw family, is ingrained into the novel as early as in Chapter IV. In that key introductory chapter, she recalls 'sitting and eating her porridge' with the Earnshaw children, and being promised 'a pocketful of apples and pears' by Old Earnshaw before he leaves for Liverpool.[309] Using this powerful scene as his initial mental template, Tilley would then find the material for the fleshing out of his adult Ellen in Marcel Pagnol's *L'Eau des collines*, which, on many points, intersected with his wife's Pyrenean family history:

> One of the most important things I did – which few others seemed to have done – was to re-evaluate the character of Nellie and work out what her agenda was. My wife, whose family home was in the French Pyrénées, brought an awareness of what the hopes of a poor village girl could have been, had she had the opportunity of working on a big farm with a lot of land attached to it. That girl could well have secretly entertained the idea of marrying the son of the house – hence the romantic attachment between Nellie and Hindley. To me, Nellie's agenda influences the way she tells the story.[310]

Fuest, the director, subscribed to his screenwriter's plan of imposing a young and compassionate type of Ellen who, if cultural accuracy were to be applied to the Brontë text, had to be impersonated by an actress in her mid-to-late twenties, the age of Ellen at the time of Cathy's death. This event, as in the Buñuel and Wyler classics, would mark the end of the translated text and Fuest, in the interview he once gave me, stressed how well his chosen Ellen served the sober *mise en scène* of Cathy's death:

> The book's first half is fire, the second half is ashes. As a story, it is one person's take on what happened – and, what do we make of her? In my version, we see Judy Cornwell with her sweetness and innocent face. But I shot the news of Cathy's death on the back of her head rather than in a close-up. Her body language told everything.[311]

The Wyler movie, which had been assimilated to the point of an 'at-homeness at the core',[312] in North America and in England, was casting its shadow on these innovative adaptative strategies, and on Tilley's 'parti pris' of translating the novel into a British period drama where socio-historical accuracy is primordial. Published in *Variety* in December 1970, the following piece by Rick Schmidt, which Schmidt believed to be relatively balanced, testifies to the sway of the old production on a North American film reviewer:

> Considering all that experience [Arkoff and Nicholson's string of low-budget period horror films] with haunted houses on hills, the selection of a rather ordinary, low level farmhouse for Wuthering Heights is curious. It may be more authentic than the set on the Goldwyn back-lot, but it is not as dramatically effective as that old, dark, windswept manor with its secret terrors.[313]

The incorporation of the Wyler-Goldwyn picture into the Imaginary of the North American critics also meant that the flawed translation of Ellen into Flora Robson – which had become a pictorial norm for the British audiences – could jeopardise the translative strategy of **compensation** by accuracy that Fuest-AIP had endorsed for the vibrant representation of Ellen as Judy Cornwell. Accordingly, in the critical pieces published when the film was released, the low neckline on the Georgian tops worn by Ellen/Judy Cornwell was often stigmatised. Was it because the more sexist critics could not properly deride her sincere, yet polished, interpretation of Ellen? Assuredly. And, was it the lavishness of Oberon's equally 'décolletées' dresses matching so well her sophisticated Hollywood persona that, thirty years earlier, spared her the comments, ranging from prudish overstatements to disparaging turns of phrase? Possibly.

In the context of the ground-breaking 1970 *Wuthering Heights*, the old sumptuary laws inherited from the feudal times, which play a major role in such a codified representational environment as the costume (or period) drama, became an aggravating factor in the stigmatisation of Judy Cornwell as Ellen. These laws do not only forbid the lady and the maid-servant to wear the same clothes or adornments, as shown in exquisite detail in Peter Webber and Olivia Hetreed's *Girl with a Pearl Earring* (2003), but also intimate that the maid should never be as pleasing to the eye as the lady she attends to. Here is David Drew's take on Tilley's Ellen, extracted from the Brontë Society's *Transactions* of 1975:

> There are also more minor alterations in the plot and various omissions and additions of characters. Lockwood does not appear in the film, the personality of Joseph is much muted, and Nellie Dean is transmuted into a frisky young girl consistently displaying large areas of bosom and apparently mildly infatuated with Hindley.[314]

Still, Schmidt, in his previously cited *Variety* review, underlines the quality of Cornwell's acting, and justifies the strategy of **compensation** by accuracy used for the representation of Ellen by its transformative effect on the spectators' gaze, which becomes sympathetic:

> If a story is to be a successful tragedy, then the audience must care about the characters.
> Only Judy Cornwell, as the buxom blonde servant through whom the story is seen and initially narrated, has that quality. The subtlety in which her love for the older brother Glover is conveyed is one of the best things in the film.[315]

Two months later, in the scathing review written by Judith Crist for the *New York Magazine*, it was the interpretation of Cathy by Anna Calder-Marshall (who, according to Schmidt, 'creates sympathy' by 'establishing her own spoiled, passion-ruled character') that came under fire. Calder-Marshall's Cathy, like Cornwell's Ellen, serves the adaptators' strategy of rendition by accuracy, a strategy that had simply not been envisioned by Hecht and MacArthur. The playwrights had directly opted for the displacement of Cathy's character into a Hollywood setting, designed for an actress who was taught, over many years, how to fit in a Hollywood movie. Crist, in a curious reversal of translative values, imposes on the 1971 readers of her review a 'betrayal by augment' as the norm for a convincing film adaptation of *Wuthering Heights*; simply because it was out of the Hollywood norm, the cultural accuracy in the representation of the 1970 Cathy felt foreign to Crist:

> She [Anna Calder-Marshall] is a girl bred in the primitive farmhouse and fields, capable of the chores at hand and the uncouth manners, simple in her admiration of refinement, wild and willful in her personal relationships. She is a far cry from the delicate exoticism of Miss Oberon, a flesh-and-blood and sensual Cathy who makes her hold on the heart believable. And perhaps it is right that in counterpoint Timothy Dalton's Heathcliff should be cloddish and brutish (ah me – we are still in thrall to the Byronic splendours of the young Olivier!)[316]

Nonetheless, below the more superficial layers of the physiques and interpretations of the actors, which were congruent with Tilley's screenplay and exceeded, in many ways, Fuest's directorial expectations, there must have been, of necessity, a translative weakness that explains the continuing disaffection of the British public for the 1970 version. More than forty years later, this original and well-crafted piece of work is still not broadcast on BBC4, which has shown so many repeats of archival dramas in recent years. It seems that, in such a codified cinema genre as the period drama hewn at home, out of an indigenous literary artefact, an over-interpretation of the source text is very much fraught with danger, even more so when this over-interpretation, or unnecessary betrayal, which is not essential to the film diegesis, can put into question the whole translative strategy of social, historical and psychological accuracy. As discussed with Tilley himself, the invented scene of the breakfast at the Heights, involving Old Earnshaw, his wife and the children (inclusive of Ellen) all sitting at the table, is remarkable for its depiction of the family tensions.[317] Mrs Earnshaw was forced to accept, as the younger son of the house an unknown seven-year-old orphan, Heathcliff, with whom her husband already prefers going to Gimmerton. Having developed a bad cough, she fears the consequences of this inexplicable attachment for the farm succession and the future of her teenage boy, Hindley. This retrospectively explains why, in a previous scene, a distraught Mrs Earnshaw chides her husband and wonders, aloud, if Heathcliff is not the result of 'his [your] own doings'. This meaningful overstatement, which signifies that Heathcliff might be Earnshaw's illegitimate son and Cathy's half-brother, was well justified by Tilley. In his long response to the criticism of a reader of *The Times*, Mrs Haussig, in April 1970, he quoted from the novel itself and referred to the interpretation of Somerset Maugham.[318] Moreover, taken in the context of the film, the dramatic impact of Mrs Earnshaw's cue helps the spectators in believing more firmly in Hindley's 'subsequent brutality and viciousness', which transforms Emily Brontë's Yorkshire tale into a very modern dark tragedy.

In another instance, however, Tilley's translative strategy of **compensation** by accuracy is compromised by a casual overstatement uttered by the adult Heathcliff. On his return from his elopement with Isabella, Heathcliff/Timothy Dalton jests about the eye colour of Cathy's unborn child and, in a crude aside to Joseph, boasts that the child could be as much his as Linton's. This casual emphasis does not blend in C.P. Sanger's (or David Daiches') timeline, and does not advance the film diegesis. It lost some critics and spectators but is consistent with the dynamic structures of the novel – the Third Bataillan Theme (Interdict 3), in particular. It was exploited by Yoshida in his own version where Kinu (Catherine) identifies herself as Onimaru's (Heathcliff's) daughter in order to avenge her mother's death. Still, in a British period drama concerned with accuracy, and where the second-generation children (baby Catherine and little Hareton) are wiped out from the screenplay before they, or their paternal lineage, can influence the course of events, this emphasis constitutes a translative mistake. In the next British period drama revisiting Emily Brontë's novel (1992), Anne Devlin and Peter Kosminsky were prompted to use caution and narratorial guidance, which often led to suppressing much of the spectators' emotional and critical response, but also has its advantages. In the instance of Heathcliff's returning to Gimmerton as a rich man, and of the six-month period he spends with Cathy (before she gives birth to her daughter and dies), Devlin resists sarcasm, and preserves the novel's tonality. Cathy/Juliette Binoche, not Heathcliff, reflects on the deadlock their story has reached by telling him, 'The surest way for you to kill me is to kiss me again'. This melodramatic overstatement brings into focus the importance of the maternal lineage and actualises the pull towards death (Interdict 3) that Cathy can already feel, while bracing the spectators for the tragic scenes that follow.

In conclusion, Tilley's attempt to work around the limitations peculiar to the format of the British period drama was highly commendable. By upgrading the character of Ellen, and translating the structural theme of Land and Lineage, he gave an extraordinary presence to the First Mythical Component (MC1) – and to the three-dimensional Gothic Figure of Lineage/*Mise en Abîme*. His emphasis on the paternal lineage that obscures, in the film, the Imaginary linked to the Second MC (MC2; Dream Visions; Childhood and the Sacred) signals a crack in his final script, not a weakness in his initial approach to the two-volume story as related by Emily Brontë. When working on the draft of his initial script, Tilley had first integrated some material belonging to the second part of the initiatory journey, but later felt compelled to withdraw it in order to stick to the exigencies of the feature film format envisaged by Fuest and AIP. Still, the important thing to remember is that Tilley had originally envisioned the theme of paternal lineage to be fully exploited, and carried into the next generation by means of a cinema sequel:

The final scene, had I been totally in control, would have shown Heathchliff lifting up the six-year old Hareton on to the table and saying: 'Now, my bonny lad, you are mine! And we'll see if one tree won't grow as crooked as another, with the same wind to twist it!'.

 I would have used that line from the novel word for word and it would have allowed for a sequel to the movie.[319]

Patrick Tilley (*Wuthering Heights*, 1970)

I recorded the interview with Patrick Tilley on a cold and sunny January morning at his farmhouse in Penyrallt, North Wales. I wish to thank Patrick for the time he gave me and the interest he showed in this interview project.

Penyrallt, Saturday 25ᵗʰ Jan, 2003

VH: Back in 1969, how did you become involved in the making of *Wuthering Heights*? Had you worked with Bob Fuest before? Did he choose you or were you headhunted by the American producers James Nicholson and Samuel Arkoff?
 Did you have the entire responsibility of the script on your shoulders?

PT: I was recruited by Peter Fullerton, assistant to Deke Heyward, producer and head of the London office of American International Pictures. I hadn't met Bob Fuest before then, but he happened to be a friend of a close friend of mine. They had been at the art college[320] together. Yes. I was given the job of producing the screenplay without any initial input from James Nicholson and Samuel Arkoff or Bob Fuest.
 Back in 1968, I had worked as production illustrator on the set of *Oh! What a Lovely War*, a feature movie based on Joan Littlewood's stage

production of Charles Chilton's play *The Long, Long Trail*. Maggie Smith, Laurence Olivier[321] and a host of other British stars all belonged to the cast. The job involved visualising the camera set-ups for the scenes that Richard Attenborough, the director, had planned to shoot.[322] But I also helped to choreograph certain scenes, for example the recruiting scene with Maggie Smith, the arrival of the Americans and the dancing scene in the ballroom.

Furthermore, during pre-production, I prepared the shooting script: it was based on Len Deighton's version of the original stage play.[323] I also wrote some additional dialogues and scenes for Olivier, Redgrave, Richardson, Bogarde and probably a couple of others I have forgotten. My name did not appear in the credits but, somehow, my work did not go completely unnoticed.

Out of the blue, American International Pictures contacted me for the adaptation of *Wuthering Heights*. It turned out that my name had been put forward by Peter Fullerton. He was at *Paramount* during the making of *Oh! What a Lovely War*.

I must say I was very reluctant to accept the job when it was offered to me. I felt that Emily Brontë's novel was part of the English literary landscape and that it was the property of what I term the 'literary mafia'. As a working-class grammar school boy, I felt I might be pilloried by my betters for daring to touch what was rightfully theirs.

Then, I thought, what the hell, why not? I read the novel five times, David Cecil's study of the novel and some other related material – including a timing of the events by David Daiches. In his foreword of the 1965 Penguin English Library Edition, he establishes the ages of the principal characters. Discovering that Heathcliff was 17 and Cathy 16, and that it was a Georgian (and not a mid-Victorian) period novel made a great difference. It helped me make sense of what turned out to be a deeply complex story. I likened it to peeling back the layers of an onion.

As for Bob Fuest, he was quite happy with my screenplay. He read it at a very early stage when it was still a long first draft counting 172 pages – as opposed to the 128 pages of the production script.

VH: How was this project of AIP born? I mean who decided to adapt once more Emily Brontë's novel?

PT: Up until 1969, American International Pictures had only produced some surf or beach movies, for instance *Beach Blanket Bingo* (1965), and some low-budget horror films. I understood that they wanted to buy themselves a certain degree of respectability in adapting a literary classic.

The initial presentation I gave a few weeks later to AIP included a time-chart of the characters and incidents, and a step outline of the novel – or perhaps that part of the story which I proposed to use in the movie. I can't quite remember. It is such a long time ago! A week or so later, Deke Heyward called me in and told me that he had given my presentation material to his thirteen-year-old daughter to read. Apparently, she had liked it and the project went ahead!

VH: You participated in the choice of the locations in West Riding, Yorkshire. Did you also have a say in the choice of the cast? (Anna Calder-Marshall, Timothy Dalton, Ian Ogilvy, Judy Cornwell)

How did the shooting go? Did you have to rewrite some of the dialogues?

PT: I did not take any casting decisions. As for Bob, it was only when he started shooting that he realised that Anna Calder-Marshall (Cathy) could not run, owing to the poliomyelitis that she had contracted as a child. Subsequently, a double was used for the scenes on the moors.

The casting of the small children (Cathy and Heathcliff) was also a bit of a problem. Besides, the teenager impersonating young Hindley was not an actor and Bob had to direct him as well as he could.

At the insistence of Anna Calder-Marshall and Timothy Dalton, the dialogues were modified to follow more closely Emily Brontë's text in the scene corresponding to the last meeting between Cathy (who is dying) and Heathcliff. I was not too happy but I suppose that they found the novel hugely involving too – once they eventually started to read it!

VH: Except for the ending which was a far cry from what you had written, were you happy with the way AIP handled the post-production and release of the movie?

Was *Wuthering Heights* successful at the box office?

PT: I truly deplore the way my ending was translated by the Hollywood producers. They took Robert Fuest's cut and re-edited the picture on their kitchen table after sending a film unit with stand-ins[324] to re-shoot the Ben Hecht tear jerker finale. Unbelievable!

But, for the first time, I had the satisfaction of seeing my name appearing in the credits, '*screenplay by Patrick Tilley*'! And after the movie had been released, AIP gave me total freedom to answer the critics of the readers of *The Times*.

At the première in London (on Leicester Square), an absolute silence fell on to the theatre when the movie ended. Can it be explained by the fact that

I had come up with a different 'take' on the novel, who knows? Anyway, between the UK and US markets the film made a few million dollars profit. Moreover, it is still shown regularly on some Swiss movie channel!

VH: You chose to delete the second half of the novel. Why?

How would you have liked the ending to be shot and edited?

The ending of the DVD version does not seem bad at all to me. Cathy's ghost leads Heathcliff up to the Heights where he is shot by Hindley. Heathcliff is then shown agonising at Penistone Crags where Cathy climbs towards him. He then literally gives up the ghost, abandoning his mortal body and yeoman clothes. Cathy and Heathcliff are shown running down the Crags. The last tableau is that of the trees, gently swung by the wind, and of the *House on the Moor*.

PT: I share David Daiches' opinion that the second half of the novel is weaker. My original ending had Heathcliff shot by Hindley, then showed Heathcliff, who was mortally wounded, still managing to ride his horse towards the graveyard and eventually falling from the saddle on to Cathy's grave.

VH: How do you think the movie was influenced by the 1970s?

And what is the place of *Wuthering Heights* in your career as a writer?

PT: In the London of the late 1960s and early 1970s, it was still fairly easy for talented young men to move from one successful artistic career to another. It did not occur to us that we could fail. My graphic designer friend and associate, Len Deighton, started to write best-seller novels (*The Ipcress File*, 1962) then to produce screenplays. The same story applies to the renowned fashion photographer, Brian Duffy, who was willing to taste what the cinema business was like at production level.

I began to pitch stories for the television and progressively put aside my graphic designer career as more film work started to come my way. After *Wuthering Heights*, AIP gave me more script assignments – one of them being an adaptation of Nathaniel Hawthorne's *The House of the Seven Gables* (1851). Unfortunately, due to the premature death of James Nicholson (1972), the partner of Sam Arkoff, the projects were dropped.

Screenwriting led me to the next important step: writing books, science-fiction based novels. I published *Fade-Out* (1975), *Mission* (1981) then the six volumes of the *Amtrak Wars* (1983-1990) and *Star Wartz* (1995).

VH: For the genesis of the script, you relied on the constructive critiques made by the likes of Somerset Maugham and David Cecil. But did you also get influenced by an artist in particular or draw some inspiration from a previous adaptation of the novel?

(The trigger for the Rivette adaptation was an exhibition in Beaubourg. He saw the original drawings that Balthus had made for the 1935 edition of *Les Hauts de Hurlevent.*)

PT: My life was transformed by the works of Marcel Pagnol[325] and the scenario of *Wuthering Heights* (also entitled *The House on the Moor*) was probably influenced by my reading of *Jean de Florette* and *Manon des Sources.*

I got a bit of a shock when I saw again the Ben Hecht movie version which, because I had absorbed so much of the book, appeared to be a total travesty of the novel with some dialogues being swapped between Hindley and Heathcliff. And, of course, the location in California bore no resemblance to the Yorkshire moors which I visited when we were looking for places to shoot.

VH: How did you contribute to the creation of such an accurate historical and social rendition of the Georgian period? On reading the script and watching the movie, it appears clearly that you were bent on describing the harshness of the existence on the moors. (As shown in the novel, Cathy's fever is contagious and deadly for both Mr and Mrs Linton, for instance.)

PT: I've always been interested in social history and did some additional research work on the period while planning the script.

VH: Do you think that someone could succeed in rendering the two halves of the novel if they adopted the principle of the *Antoine Doinel* serial of François Truffaut which spanned twenty years of the life of its character (1959-1979)?

Also, you put some unusual depth in Edgar's character. Is it your way of acknowledging the second part of the novel?

PT: The final scene, had I been totally in control, would have shown Heathchliff lifting up the six-year-old Hareton on to the table and saying: 'Now, my bonny lad, you are mine! And we'll see if one tree won't grow as crooked as another, with the same wind to twist it!'.

I would have used that line from the novel word for word and it would have allowed for a sequel to the movie.

So the principle of the Antoine Doinel series could have worked.

As for Edgar, his character evolved for the better and became less 'girlish' in the process of the screenplay maturation from the original to the production script.

VH: From the first to the last page of the script, you managed to put some good logic into the characters' motivations and resulting actions. The incest motif and the questionings about Heathcliff's origins are constantly brewing under the surface of the novel and you merely brought them into the open in your version. Therefore, I partly disagree with Mrs Hanna Taussig who complained in writing about your bold assumptions (*The Times*, 15th Apr, 1970).

In your script, we know from the start that Heathcliff could well be the result of 'Mr Earnshaw's own doings'. Indeed, the latter clearly favours Heathcliff when taking him, and not Hindley, to Gimmerton – an episode which is not present in the novel.

Similarly, hinting that Cathy's baby might well be the child of Heathcliff (in the aside[326] taking place between Joseph and Heathcliff after Isabella's elopement) puts the accent on the new connivance between old servant and new master, and is in line with the rest of the script.

Is it to put into perspective the rigid social structure of Georgian England that you interpreted the way you did the romantic attachment between Nellie and Hindley?

PT: One of the most important things I did – which few others seemed to have done – was to re-evaluate the character of Nellie and work out what her agenda was. My wife, whose family home was in the French Pyrénées, brought an awareness of what the hopes of a poor village girl could have been, had she had the opportunity of working on a big farm with a lot of land attached to it. That girl could well have secretly entertained the idea of marrying the son of the house – hence the romantic attachment between Nellie and Hindley. To me, Nellie's agenda influences the way she tells the story.

VH: The scene in which Cathy, clad in her grand dress, prepares her horse for a ride to the Grange, only to be thrown on to the stable floor by Heathcliff, announces the scene of love making in the park of the Grange. How was the latter scene received by the critics?

PT: As far as I know, they did not make any comments on that. I had visualised the scene in the park as coming immediately before the appearance of Heathcliff at the Grange after his three-year absence. It would have given the actors even more emotions to play with during the late afternoon tea party but Bob Fuest thought it would have gone too far.

VH: What about the visual recurrence of the flower motif, be it harebells, bluebells or crocuses? Was it part of your original script?

PT: No. The credit for the way the film looks must go to Bob Fuest. He may have drawn some pictures from the script but he was trained as a painter and had a wonderful visual imagination.

VH: As to the transitions, did they come naturally to you? I am alluding, for instance, to Isabella's elopement as announced by the amorous porcelain figurine 'à la Choderlos de Laclos' and to the wood panel of Frances' coffin merging into the gambling kitchen table.

PT: Those two particular transitions are part of the editing process which the director controls.

VH: The monologue of Cathy ('I am Heathcliff') is remarkable in efficiency owing to the acting qualities of Anna Calder-Marshall but also to the scenario. The scene is split between the indoors (the kitchen) and the outdoors (the farm courtyard). Was it part of your original script?

PT: Yes, it was. And I still believe my screenplay was the closest rendition yet to the book.

Robert Fuest (*Wuthering Heights*, 1970)

'Wuthering and Me'

I would like to thank Bob and Jane Fuest for their time, their good humour and their kind hospitality.

Winchester, 25th May, 2003

VH: Back in 1969, how did you happen to get chosen for directing *Wuthering Heights*?

Had you already worked for Samuel Arkoff and James Nicholson, the producers of American International Pictures? Were they familiar with your work as director on *The Avengers*[327] or with your excellent contribution to British television in general?

How was *Wuthering Heights* received by the critics?

BF: I owe my introduction to AIP to Bryan Forbes,[328] a writer-director who had just taken control of the studios at Elstree. I had just finished *And Soon the Darkness*[329] on which he was executive producer. (For the record, the first feature film I directed was *Just Like a Woman* (1968). It was a long and protracted story taking place in the swinging 1960s but it did win a prize at the Edinburgh Film Festival in 1968.)

So my second feature, *And Soon the Darkness*, was a modest thriller shot in Orléans and in the studios at Elstree. Bryan Forbes was more than a figurehead wearing hunting jackets. He had charisma, could act, write and direct. He commissioned five or six feature films on taking over Elstree, announcing them as participating in some sort of British film renaissance – which was a mistake. Only two survived, my own and *Cry of the Penguins*[330] starring John Hurt as Forbush, a young biologist from London.

And Soon the Darkness still survives but the critics were very antagonistic at the time. I must say after seeing it at a recent festival[331] held by the Manchester Film Studies Institute that I think it really deserved more. We were within an inch of getting Bernard Herrmann[332] to do the score and he did say – although he was not a man of compromise – that 'Hitchcock would have been proud'. (The fact that I think that Hitchcock is vastly overrated does not change what he said then.) With *Wuthering Heights*, it is the exact same. There is much more appreciation for it now than there was before.

Anyway, I brought *And Soon the Darkness* well within budget and Bryan Forbes was kind enough to mention me to AIP. Samuel Arkoff and James Nicholson were an extraordinary coupling. Stories abound on how they first started but I believe Arkoff worked in the local cinema and Nicholson was the projectionist. Somehow they dragged themselves out and into the area of production. Most of their films were Z movies – Beach flicks and Drive-in movies. They had a terrible reputation and yet the industry was in awe of them. At their peak, they had an advertising thrust that was laughable but enviable. One lunchtime, I remember walking with a member of the William Morris agency. He pointed to a pile of dog mess on the curb and said: 'Give them a week and AIP would make a fortune out of that'.

I only mention this because the aura of cheap exploitation always prevailed with AIP. In retrospect, I think it influenced the reviews of *Wuthering Heights*. It certainly did in the UK where the film was almost universally condemned. It is interesting how times and attitudes have changed but back then it was painful. We all gave it our best and all felt the wounds with such headlines as 'Carry on Wuthering' and 'Wuthering Depths'[333] in the most intelligent broadsheets. Now the critics have changed their minds but, then, we were all dreadfully hurt.

VH: How was this AIP project born? Were you happy with the way AIP handled the production of the movie? How did the shoot go since it was only, after all, your third feature film? Did you have a say in the choice of the cast? (Anna Calder-Marshall, Timothy Dalton, Ian Ogilvy, Judy Cornwell)

BF: It was a very happy picture but it was a very hard film to make on schedule. I remember the cold of the Yorkshire moor especially. Thermal underwear was unknown at the time: the crew and cast were all given some as a try-out. It was marvellous to work without being completely frozen!

Neither Sam nor Jim had read the book, but they had seen the Olivier B/W movie and read the Educational Classic Comic Book version – which only includes the first part of the book. They both felt sincerely that they deserved some better recognition and the prospect of *Wuthering Heights* getting a Royal Command performance was very tempting – and they got an official reward in the end, for some obscure reason, a medal from the Pope!

In all honesty, I have to say that once agreeing upon making the film they did not interfere or attempt to prejudice me in any way. They wanted Olivier to play the father (but it did not happen) and they liked Tim Dalton because he looked like the young Olivier. Anna Calder-Marshall, 'Cathy', was perfect for the part. She had just done a TV series[334] which, at the time, was of special interest. As I recall it, it consisted of some separate one-hour dramas, each with a different male lead in the same league as Connery and Caine.

Working with Tim Dalton and Anna Calder-Marshall was a delight. Anna was extraordinary, wonderful. One rehearsal and somehow they enacted a chemistry which would have been criminal to disrupt, a tension (even in the more passive scenes) I likened to a cat's cradle. For each of them was responding to the other quite intuitively. There was a certain roughness about Tim then – later, being James Bond cooled it. The rest of the casting was left to me – as was the AIP crew who had worked previously on the UK films. I am afraid I had to fire two of them.

VH: What is your opinion on the novel? When did you read it for the first time? What about the Olivier movie directed by Wyler in 1939 and also the five-episode TV serial, *Wuthering Heights*, directed by another big name of *The Avengers*, Peter Hammond, in 1978?

BF: I read *Wuthering Heights* when I was at school and not since. Of all the great classics, I think it is the most un-filmable. The book's first half is fire, the second half is ashes. As a story, it is one person's take on what happened –

and, what do we make of her? In my version, we see Judy Cornwell with her sweetness and innocent face. But I shot the news of Cathy's death on the back of her head rather than in a close-up. Her body language told everything.

I thought that the B/W Hollywood classic version quite pathetic. After getting the job, I ran it and was amazed how awful it was – but perhaps if I saw it now…

I have not seen the 1978 TV serial but when I worked for ABC/Thames Television, my hero was Peter Hammond. As a production designer, I worked with him on the first episodes of *The Avengers*[335] that he directed in 1961. Seeing him work was one of the inspirations that made me consider directing. I would have loved to see his *Wuthering Heights* but I was in New York at the time. The fact that he made just one feature film[336] was in itself rather disappointing.

VH: To come back to your *Wuthering Heights*, how long were you given to shoot the movie? Did you choose yourself the location in West Riding, Yorkshire? Which studios did you use?

BF: We shot in seven weeks, three on location – shooting with horses is time consuming, especially when you need several takes! – and four at Pinewood Studios. The location was outside Leeds – I'm afraid I cannot remember where. But it was very hard to find and there are quite a few funny stories attached to this venture. The art director, Philip Harrison, and I spent about a month looking for the right locations. Even then, it was hard to find some moorland without any distant electric pylons. Also, we needed parking space for the numerous cars of the extras. Philip Harrison had worked with me at Thames Television. Because the moor was so exposed, the line producer from AIP, Louis (Deke) Heyward, only came on the first day of shooting. But he was a splendid man – a terrific character – who sadly died last year.

VH: How was the première of *Wuthering Heights* in London? What about the reception of the film in New York?

BF: Jim and Sam only appeared at the London première. A Royal Gala was all they wanted. Despite the lousy reviews, the film did not do too badly here and soon made its money in the States, playing for six weeks at Radio City.

VH: The score of Michel Legrand reminds me of the music he wrote for Jacques Demy's *Peau d'âne* starring Catherine Deneuve, especially when Nellie starts recounting, in the true accents of a Fairy Tale, the arrival of Mr Earnshaw at the Heights with Heathcliff buried under his coat.

BF: I do not know this movie. At the time, Michel Legrand used to live in a lovely water mill, not far from Paris, and had this pony and trap to gallivant around the countryside. I stayed with him for a few days. One night, after we had had a few drinks, we recorded a cassette with him singing and playing the piano. In retrospect, there is too much music and not enough silence in *Wuthering Heights*. There is too much Strauss and not enough Bach, if you prefer. Bach is thinner and sparer, and gives more tension with fewer musicians.

VH: How would you have liked the end to be shot and edited? The ending that I know from the DVD version of the movie can be summed up this way. Cathy's ghost leads Heathcliff up to the Heights where he is shot by Hindley. Heathcliff is then shown agonising at Penistone Crags where Cathy climbs towards him. He then literally gives up the ghost, abandoning his mortal body and yeoman clothes. Cathy and Heathcliff are shown running down the Crags as in the 1939 version. The last tableau is that of the trees, gently swung by the wind, and of the *House on the Moor*.

BF: There was a problem with the ending. It had to be invented. I wanted a shot of Dalton's dead hand with the lace cuff in the wind on Penistone rock. The issue was that it was not a happy ending. At first, I refused to change it. However, since they would have made some stand-ins do the job anyway, Tim, Anna and I eventually agreed to shoot a few more scenes and give it a happy ending. Funnily enough, when I took a copy of the film to the Netherlands and said I did not like the ending, an old projectionist with a cigarette stub hanging out of his mouth cut the print here and now, in the projection box, with a simple pair of scissors and stuck back the credits at the end!

Kosminsky-Devlin (1992, Paramount Pictures, 105 min) and Skynner-McKay (1998, ITV1, 112 min)

The legacy that the 1970 *Wuthering Heights* bequeathed to the British period dramas of the 1990s is ambivalent. This legacy affects predominantly the socio-historical backdrop berthing the First MC (MC1), and the conventional choice of a British, then Scottish actor (Ralph Fiennes, in 1992, and Robert Cavanah, in 1998) for the role of the 'dark-skinned gypsy' Heathcliff.[337] In both productions, however, the narrative structure is much altered since it encompasses the two generations of lovers for a comparable film duration of 105 minutes in 1992, and for a screen time only increased by seven minutes in 1998.

In the 1992 version by Kosminsky and Devlin, the First MC (MC1) is not set on the same representational scale as in the 1970 period drama. If a similar vision translates for the big screen the daily hardships of farm life in Georgian Yorkshire, as well as actualises the divide between yeomen and servants, the characters inhabiting, in the words of Lucasta Miller, the 'grandiose Gothic edifice' called Wuthering Heights, or sitting on the sea of stones subsuming Penistone Crags, are often dwarfed by the settings. In other words, the uncompromising heroes interpreted by Juliette Binoche and Ralph Fiennes do not seem to be potent enough to impress the tragic events that they experience on to the grand, unmovable stones which surround them:

> The opening sequence has Emily Brontë (Sinead O'Connor) exploring a now derelict Wuthering Heights, no longer the rough farmhouse of the original, but a grandiose Gothic edifice. […]
>
> Less audacious, Kosminsky's film shadows its source more closely in externals, but the result is less satisfying. […]
>
> The quest for visual authenticity ends up cramping the story's emotional impact.[338]

By reinforcing their First MC (MC1) with a *mise en scène* and script reminiscent of Buñuel's *Abismos de pasión*, Skynner and McKay, in their 1998 version, avoid this 'stony' hurdle and come closer to the scale and register that were fine-tuned by Fuest and Tilley in their 1970 production. In the interview he gave me, director David Skynner and I discuss how the line between the animate and the inanimate were blurred. On the one hand, Skynner works at magnifying the imprint of farm life on the buildings (and on the land) of the Heights and, on the other, at endowing the farm (pig, sheep, lamb) and wild (crow and fox) animals with some contrasting metaphorical meanings. Skynner's poetical aesthetics combined with his realistic type of *mise en scène* did not only make tangible the decay that takes hold of the farm under Hindley's rule, but also showcased the renewal of life that pushes abjection inexorably away; simultaneously, McKay's narrative arc pointed all along towards Heathcliff undergoing his final transformation, and Catherine and Hareton acknowledging their feelings for one another:

> You mentioned the sheep but there is a little lamb standing next to the dead sheep – which, by the way, was not real – as Lockwood is coming up to the Heights. This exemplifies very well your idea of visual metaphor. Here is death but new life is struggling to escape it.[339]

Skynner's interview, reproduced over the next few pages, also highlights the effort of Coleridgean suspension of disbelief that the film viewers of *Wuthering Heights* are required to make in order to perceive as 'authentic' not only the settings and costumes, but also the impersonations of the actors – in his version, Orla Brady and Robert Cavanah, as Cathy and Heathcliff, and Sarah Smart and Matthew MacFadyen, as Catherine and Hareton.[340] As evoked in our previous Chapter dedicated to the analysis of the BBC teleplays and *Classic Serials*, the audience response to the rendition of a classic novel into a staged period drama is conditioned by casting, acting and *mise en scène*. With *Wuthering Heights*, this effort is rendered strenuous by the fact that the source novel is a condensed family saga staging three generations, and involving two radically different families. Those two families possess some specific physical and mental traits that reflect symbolically on their environment, and stimulate the action. Heathcliff and Ellen, the sole characters who evolve through all three generations, and share the intermediate status of adopted family members, would ideally be mirrored by the casting, on screen, of two young adults since Heathcliff only enters his late thirties, and Ellen her mid-forties, in the second half of the story. The casting of a 'youngish' Heathcliff would only materialise with the casting of Ian McShane (1967) and Timothy Dalton (1970), while the impersonation of a younger Nellie was favoured in the 1967 *Classic Serial* (and in the 1970 and 1992 feature films), but not in the 1998 television movie.

In any case, there still remains the very problematic casting issue of Heathcliff as an adolescent in his early teens, which only the lost movie by Albert V. Bramble (1920) and the most recent feature film by Andrea Arnold (2011) have so far tackled. Sara Martín, in 'What does Heathcliff Look Like? Performance in Peter Kosminsky's version of Emily Brontë's *Wuthering Heights*' (2005), voices her concern about his representational authenticity as a teenager, and articulates the paradoxes of narrative (or translative) fidelity within the genre of the period drama at large:

> [...] child actors are cast for the initial scenes but Olivier, Dalton and Fiennes are made to face the impossible task of playing teenage Heathcliff. This is easier for Dalton, aged just 24 when he played the role, but Olivier (32) and Fiennes (30) look plainly too old for the scenes covering Heathcliff's adolescence [...]. The problems concerning this aspect of the adaptation depend thus on the inability to simultaneously maintain and ignore narrative fidelity.[341]

In 1991-1992, authenticity in representation and performance was at the centre of the preoccupations of British director, Peter Kosminsky. For his Hollywood-commissioned, 'made in Britain' version, he was asked by the American studios, Paramount Pictures, to work with charismatic actors Ralph Fiennes, the pride of the Royal Shakespeare Company, who had appeared as T.E. Lawrence in a high-quality television remake of *Lawrence of Arabia* in 1990, and Juliette Binoche, a rising film star, who had received European acclaim for her prominent parts in *The Unbearable Lightness of Being* in 1988, and *Les amants du Pont-Neuf* in 1991. In this context, Kosminsky was facing the arduous task of appealing to a wide-ranging public of fans, while his actual mission was to achieve performance authenticity in his *mise en scène* of *Wuthering Heights*.

The combination of a talented actress speaking with a French accent and a handsome Shakespearean actor, untried in his new Brontëan role, was audacious, but meant trusting in Fiennes' and Binoche's international statures. They brought a certain degree of spontaneity and untameness into their characterisations of Cathy and Heathcliff, prefiguring their remarkable performances in *The English Patient*, four years later. In *Wuthering Heights*, nonetheless, Fiennes and Binoche were not able to shake off their idiosyncrasies (and recent acting pasts) as convincingly as in Anthony Minghella's powerful rendition of Michael Ondaatje's novel (1996). Furthermore, Kosminsky could not question Paramount Pictures' commercial tactics, which not only required Binoche's successive incarnations of a dark-haired Cathy Eanshaw then a fair-haired Catherine Linton, comparable with Angela Scoular's impersonations of both mother and daughter in the 1967 *Classic Serial*, but also dictated the photogenic interpretation of the fictional narrator, Emily Brontë, by pop singer Sinéad O'Connor:

> Performance, on the other hand, is also conditioned by the intertextual contribution that casting makes to a particular role, that is, by the effect that the actor's looks, career and cast companions have beyond the adequacy or proficiency of the actual performance.[342]
>
> (Sara Martín)

> Peter Kosminsky's 1992 adaptation [...] keeps its distance by introducing a female narrator as an intermediary. This surrogate for Emily Brontë wanders through the ruins of the wind-blasted farmhouse as she tells the story and implores us 'not to smile at any part of it'. If we need to be advised against sceptical tittering, the battle is already lost.[343]
>
> (Peter Conrad)

Fortunately, the 1992 and 1998 period dramas did not limit themselves to representational or performance authenticity. As in the innovative 1978 *Classic Serial* by Snodin and Hammond, they sought to increase the level of translative accuracy by acknowledging the *Mise en Abîme* architecture of the novel and by staging, albeit more briefly, Lockwood's Self-Revelatory (Romantic) Journey. Devlin, in her 1991 script, set out to integrate Cathy's box-bed into the film narrative and transform it into a refuge as significant, for Cathy and Heathcliff, as the cave of Penistone Crags – upgrading it into a symbol that reproduced the three-dimensional architecture of *Mise en Abîme*, inside the Gothic mansion itself. The film, however, fails to convey all the emotional charge intended for the spectators. The authoritative insistence of the camera on the grandeur of the sets ends up drowning the carefully woven motif of the box-bed and, in a borrowing from Kiju Yoshida *Ozu's Anti-Cinema*, prevents the spectators from 'obtain[ing] a comparative perspective that allows them to think about the differences between the two films, the film they are watching and the film the characters are watching'.[344]

Further, the unifying gaze and controlling voice of the staged intra-diegetic author, Emily Brontë, appear to deeply modify the *Mise en Abîme* Figure and its propensity to free characters (and film sets) from an omniscient narrator as well as offer – in Yoshida's words again – some 'dynamic multilateral perspectives' to the spectators.[345] This overbearing intra-diegetic author appears to be, together with a blatant overstatement from the extra-diegetic author – I am referring to Tilley's misplaced emphasis on the eye colour of the unborn baby in his 1969 script – the major pitfall of this translative strategy of compensation by ***accuracy***, which is favoured in all three of these British period dramas (1970, 1992 and 1998).

In her previously cited essay on the 1992 version, Sara Martín regarded the *appropriation* of the Gothic Figure by the intra-diegetic author as a 'narratorial' flaw – as opposed to a 'dramatic' flaw, linked to the performance of the actors. This *appropriation* of the *Mise en Abîme* decreases the levels of translative fidelity by limiting the possibilities of trespassing and learning of the intra-diegetic outsiders, and their exploration of the Bataillan Themes (BTs):

> [...] the narrative framework created in the film by introducing Brontë is a poor substitute for the complex narrative frame of the novel and, what is more, contradicts the author's decision to erase herself from the text and use Nelly Dean and Mr Lockwood as (possibly unreliable) narrators. O'Connor's presence 'authorises' the film in a way the novel specifically avoids, offering as the truth what in the novel is possibly just a version of the events.[346]

* * *

If Lockwood was to be all together jettisoned by Peter Bowker in his script (2008), his impact as a character 'en devenir' had already been – as we have just seen – considerably diminished by Anne Devlin (1991) who imposed the presence on screen of Sinéad O'Connor/Emily Brontë, the intra-diegetic author. The 1998 period drama directed by David Skynner and written by Neil McKay featured prominently Lockwood, the intra-diegetic outsider. Like Peter Hammond's version, David Skynner's version relied on a strong cast for the second generation of lovers, Matthew MacFadyen, the future Darcy of Joe Wright's *Pride and Prejudice* (2005), as Hareton, and Sarah Smart, the future Carol Bolton/Heathcliff of Sally Wainwright's *Sparkhouse* (2002), as Catherine. These strategic casting choices increased the levels of both narratorial and dramatic authenticity, and made the second part of the Initiatory Love Journey (from Thanatos to Eros) memorable.

The 1998 television production was a successful venture; on the first evening of its broadcast (5th April), it brought ITV1 a third of all the audience ratings. This genuine success, when related to the corpus of adaptations of *Wuthering Heights*, substantiated Nigel Kneale's prophetic views on the interrelation between television and cinema, which were already four decades old. In a *Sight and Sound* article famously entitled 'Not Quite So Intimate' (1959), Kneale had envisioned that 'television drama at its best' would become 'almost identical with film at its best', and vice versa.[347] Commissioned by ITV twenty years after David Snodin's next-to-perfect adapted script for the 1978 *Classic Serial*, which marked the end of an elaborate, purely televisual style that was admirably mastered by Peter Hammond, the 1998 television production would be embracing the *Classic Serial*'s costume (or period) drama genre, while leaving behind the serial format for a translation of the novel into a feature film. Despite appearances, the following adaptation, the costly looking 2009 production written by Peter Bowker and broadcast in two parts on ITV1 (30th and 31st Aug), was not designed as a serial either. It was a lengthy television movie (137 minutes) spliced with a conspicuous string of advertisements (43 minutes), not a two-part serial lasting 180 minutes in total, as it was then advertised by the channel.

Back in 1998, the first television movie of *Wuthering Heights* could boast a balanced translative strategy, both narratorial and dramatic, which the 2009 television movie might have arguably been lacking in. In retrospect, the 1998 version had been compromised by a missed opportunity: letting an excellent British actor of Indian descent, Naveen Andrews ('Kip' in *The English Patient*), play the role of Heathcliff. As explained by Skynner himself in his interview, Andrews would have been his first choice, had the producers' racial bias been negotiable:

> Heathcliff is well described in the novel as being a dark-skinned character[348] – who could be of East Indian descent, probably the son of an Indian sailor. But the simple fact that he was on the list created a smouldering opposition. A lot of people at LWT – Sally included – said 'we are not sure about that, we are not having him yet'. I met him though and was just blown away by his intensity. He really struck me as a young woman's hero, intense and fierce. He wanted to know why, in the script, there were no references to him as being Indian – of course, we had not had a chance to adapt it by then.[349]

Skynner's account confers an empirical basis to Sara Martín's sharp discussion on the limits of performance authenticity in 'What Does Heathcliff Look Like?' (2005), a sharp discussion already emphatically quoted. There, Martín also underlines the lack of authenticity in the 'racial' representation of Heathcliff, even in the most recent British period dramas. She simultaneously asserts the existence of an 'ideal' representation, yet to be attained, to gain in translative fidelity:

> By casting actors who are quite different from Brontë's dark, gypsy-looking Heathcliff, the films simultaneously expose their own racism and the very limits of fidelity, perhaps also conditioned by the reader's difficulties to visualise Heathcliff, as Emily Brontë did [...] Given the importance of racial issues in our time, this might well turn out to be the main focus of a possible future adaptation of *Wuthering Heights*.[350]

Peter Kosminsky (*Wuthering Heights*, 1992)

Written answer, 19ᵗʰ Jan, 2005

VH: When it comes to conveying a sense of clarity to an audience, mixing biography and fiction and giving the author a voice and a face[351] can be seen as taking a little bit of a risk.[352]

Did you feel comfortable with the fact that Ileen Maisel (the representative from Paramount Europe) and Mary Selway[353] (the producer) decided from the very beginning that the 'strong presence' of Emily Brontë 'should be reflected in the film'.[354]

Was her fictional presence imposed upon you in the opening and final sequences?

PK: It isn't fair to say that they imposed it upon me. I agreed with the strategy. In fact, it may even have been my idea.

VH: The character of Emily Brontë is also silhouetted on the moors to introduce swiftly the flashback that recounts Heathcliff's arrival at the Heights with the old Earnshaw.

Was her character originally meant to appear more often on the screen?

PK: It is certainly true that I wanted Emily to appear when a remark from her (or an indication that she was thinking hard about the moment she was depicting) would be helpful.

VH: Was her voice-over always regarded as the most efficient means of keeping the audience on the narrative track – just in case anybody got lost?

Also, would you like to tell me about the making of Lockwood's nightmare scenes? These two complementary scenes – one seen from the outsider's viewpoint (Lockwood), the other seen from the insiders' viewpoints (Heathcliff, Catherine and Hareton) – frame the beginning and the end of Sinéad O'Connor/Emily Brontë's recounting of *Wuthering Heights*. Somehow, the screenwriter[355] and you succeeded in re-creating the 'trompe l'oeil' perspective of the novel's embedded narratives and allowed yourselves to be guided by the Gothic blend of realistic sensations and hallucinatory visions that is present in the novel.

PK: The voice-over was helpful to navigate the story. We were confined to a 105-minute format – and to tell both halves of the story in that timeframe was very, very difficult. We did use her v/o to jump us forward from time to time. As far as Lockwood's nightmare is concerned, we tried to faithfully depict the scene as written by Emily in the novel.

VH: How did the orientation of Anne Devlin's second draft change when you joined her to work on the script around December 1990.[356]
Were you both happy with the final (and fifth) draft that, I believe, was completed in June 1991? [357]

PK: The basic thrust of the script was in existence before I joined. Of course, certain changes were made as a result of my involvement. I think it is probably true to say that I was keen to re-insert more of the original dialogue from the novel. But, in essentials, the script didn't change.

VH: Was it then the first time that you got involved into the writing of a script?

PK: Not really. I had been a script editor at the BBC previously and had collaborated extensively with Michael Eaton on the script for *Shoot to Kill* in 1990.

VH: How did your work on *Wuthering Heights* differ from your previous experience as a director on television documentaries? [358]

PK: The scale of the undertaking, the sums of money involved, the involvement of an American studio, the need for well-known actors, the relentless commercial drive, the expertise of the crew we were able to attract (being a movie) and the fact that *WH* was a period piece – something I had not done before (or since).

VH: How did you get on with the actors on this shoot?

PK: I think we all got on extremely well. I have very fond memories of those collaborations.

VH: Do you have any comments on the ridiculous copyright issue that was raised by the Goldwyn Company[359] over the use by Paramount of the title of the Samuel Goldwyn movie, *Wuthering Heights* (1939)?

PK: I'm afraid not. In fact, I had forgotten about this until you mentioned it. I don't remember it as being seen as a particular problem at the time.

VH: There is a deep connection between your movie and Bob Fuest and Patrick Tilley's *Wuthering Heights* (1970) despite the fact that they only adapted the first part of the novel. In both movies, the socio-historical background is made palpable through an actualisation of the harsh living conditions on a Yorkshire farm in the Georgian times and an accentuation of the social divide between yeomen (or landowners) and servants. Would you like to comment on any of the following sequences?

Young Cathy standing alone in the kitchen and savouring quietly the bowl of milk brought in by Nellie on that freezing winter morning while Heathcliff, and the other servants, who have been up since dawn, are busying themselves outside with the farm's chores.[360]

Baby Hareton's christening with the gentry (the Earnshaws and the Lintons) standing on each side of the priest, in the light of the window, and the servants and labourers standing at a respectable distance, in a dark corner of the room.

The grave sequence with the prolonged framing of Cathy's headstone (accompanied by the priest's sermon in a voice-over) followed by a slow camera pan on the more modest tombs of the graveyard, which ends up with a long shot of Hindley's burial ceremony, six months later (accompanied by Sinéad O'Connor's voice-over).[361]

PK: I'm just glad you liked them. I do remember fighting hard to keep in the grave sequence (your third point above), which I especially liked. I think when the execs first saw the shot they told me that I was stupid to assume that a shot of that length could be included and asked why I shot it. But it crept back into the film towards the end of the edit and I, for one, was very glad.

VH: What pushed you and Anne to postpone Heathcliff and Cathy's last encounter until after the birth of her baby girl?

In the novel (and most of the other film adaptations), the intensity of this last encounter appears to trigger Cathy's labour pains and leads her to her grave. I have the feeling that you might have chosen to bring out the one similarity behind Frances' and Cathy's untimely deaths: giving birth in Georgian times.

PK: I'm really sorry, I just can't remember. It's so long ago. I imagine it was because it helped bring the story to some kind of climax but anything more I say would be just invention.

David Skynner (*Wuthering Heights,* 1998)

On a bright February afternoon, I met the film director David Skynner at the British Library in London. I wish to thank him for his kind interest in the discussion and for the time he gave me.

The Readers' Lounge of the British Library, 16ᵗʰ Feb, 2005

VH: Back in 1997, how did you become involved in the making of the first single-episode version of *Wuthering Heights* on television (ITV)?

Did your early cinema experience as a third assistant director on *A Fish Called Wanda* (1988) encourage Granada/London Weekend Television (LWT) to recruit you?

Or was it rather your directorial success in 1996 with the popular television drama series, *London Bridge,* on ITV that caught their attention?

How did it all start?

DS: Neither *London Bridge* nor *A Fish Called Wanda* – on which I was an assistant director many, many years ago! – gave me the opportunity to embark on this film.

After *London Bridge,* I did a series on ITV called *Staying Alive* (1996) which was a sort of thriller set in a nurses' home. The executive producer for Granada/LWT was someone called Sally Head, who was very well known as the producer of some ground-breaking, significant drama[362] – and became an executive producer early in her career.[363] When we finished *Staying Alive,* I met her at the after-shoot party and she asked me if I would be interested in doing *Wuthering Heights*: she was about to put together the budget for it. And I said 'Yes'. I was the first person she chose, the writer and the producer came later.

It [*Wuthering Heights*] was not a particularly happy job for me. There were a lot of personality conflicts, largely with the producer, Louise Berridge, and the director of photography, Peter Middleton *BSC* – with whom I managed to work well anyway. Therefore, I was invited by Sally Head. This is really how I came to be involved in *Wuthering Heights*.[364]

On the one hand, I had my doubts about making the film when I understood who was going to be producing it. On the other hand, I had already met the writer, Neil McKay,[365] before on *Staying Alive*. So I stuck to the project.

VH: Did you feel that Louise Berridge, the producer, was not giving you enough leeway?

248

DS: Louise, who comes from an academic background, had a very rigid idea of what she wanted to do.[366] She knew that the production would be looked at with a very critical eye in some quarters. She could be quite heavy-handed when she wanted to make sure that it was done her way.

VH: Did you collaborate with Neil McKay on the script of *Wuthering Heights*? If not, could you tell in which pre-production areas you took part?

Did you see *Sparkhouse* which was broadcasted on 1st and 8th Sept, 2002, on BBC One?

Sarah Smart, your sweet second-generation Catherine, was cast as Carol Bolton (alias Heathcliff), the daughter of a brutal Yorkshire farmer. Sarah impersonated 'a female character who was [created to be] as wild and angry and screwed up, but also as magnificent and passionately in love as Heathcliff' – after Sally Wainwright (30th Oct, 2002).

DS: Neil wrote on his own. I did not have a lot of input when it came to the script. I wanted to, but was not allowed to for a number of reasons. In everything else, from the casting of the actors to the directing of the film itself, it was completely different.

There was a bit of a battle in pre-production between Sally (Head) who, as in the 1970 version, only wanted to do the first part of the book and Louise who, rightly I thought, wanted to do the full story. To me, it was precisely what was interesting, making the full story work as best as we could. Structurally, it is a very difficult story to adapt, a nightmare! I do not think anyone got it right. I have not seen *Sparkhouse*: I have just heard good things about it. But, in taking the story away from the actual characters of Cathy and Heathcliff, in effecting a reversal of sex, I think that, maybe, Sally (Wainwright) may actually have taken the curse off it. There is a real curse floating around it! It is a quasi-impossible story to adapt, being very melodramatic in a quite problematic way. In pre-production, it became quite a political battle to decide whether it was going to be half the story or the whole story. In the end, from an artistic point of view, it had to be the whole story.

VH: Back in 1991, Anne Devlin had managed to write a script including the two halves of the novel for Paramount British Pictures. With this 105-minute version directed by Peter Kosminsky (and starring Juliette Binoche and Ralph Fiennes), Neil McKay would have had a condensed cinematic example to look at.

Were Sally and Louise able to take this chance to discuss with Neil what they wished he could keep or what they wanted him to avoid?

DS: This, I cannot tell. All I can say is that it is two different stories and virtually two books. As you know, up to the point where Cathy dies, it is one book and then the succeeding generation is another book. As a consequence, the story seems to have several endings: the novel keeps ending and ending… It is almost impossible to get a strong narrative arch that carries you from beginning to end: you do fall flat after Cathy's death.

To me, though, the most successful part of my production was the second half. I am not happy at all with the first half!

VH: I personally found that the casting of Sarah Smart and Matthew MacFadyen as Catherine and Hareton was an excellent idea. They work very well together on the screen and their youth fits perfectly the ages of the characters of the novel. Also, their respective performances feel fresh and natural. Unlike Orla Brady and Robert Cavanah, who were then already much older than Emily Brontë's Cathy and Heathcliff, they probably felt freer from the grand expectations of the press and of the audience.

DS: Matthew MacFadyen and Sarah Smart are deeply charming, very sensitive. I think that there are some very beautiful scenes between them. I love the scene where he is reading poetry and she is teasing him – this is probably my favourite scene in the whole film – and then they shut up as Heathcliff walks past them. It was a lovely day, a beautiful summer day. She was planting her flower garden, he was reading poetry. A very beautiful, moving scene…

I was very happy when I found Matthew and Sarah. They have gone on to greater and greater things. Matthew is now Darcy in *Pride and Prejudice* (2005, Working Title Films). He is very successful.[367] He has 'major film star' written all over him – and always has had.

For the first half, it was very hard to find people: we had a list of eight cast actors. At that early stage, I used to like a lot the idea of Jude Law. Having seen much more of his work now, I do not think he could have carried it though. But it was an interesting idea. What happened – because there had been so many unsuccessful productions before – is that a lot of people went 'Oh, no! I am not interested…' and dropped out, without reading the script.

VH: Is it true that Colin Firth and Ewan McGregor were asked to play the part of Heathcliff?

DS: Colin Firth.[368] I expect he was on my list… He would have been quite good. But I am sure he ruled himself out: I never heard of him. As to Ewan McGregor, he had already appeared in too many films and had probably left that orbit. I think he was doing *Star Wars* at that point. He is a very, very busy guy.

But I have one regret actually. We did the screen tests of a short list of actors and one of them did not turn up, an actor called Naveen Andrews,[369] who is Indian.

VH: He could perfectly have played Heathcliff!

DS: He would have been perfect! Heathcliff is well described in the novel as being a dark-skinned character[370] – who could be of East Indian descent, probably the son of an Indian sailor. But the simple fact that he was on the list created a smouldering opposition. A lot of people at LWT – Sally included – said 'we are not sure about that, we are not having him yet'. I met him though and was just blown away by his intensity. He really struck me as a young woman's hero, intense and fierce. He wanted to know why, in the script, there were no references to him as being Indian – of course, we had not had a chance to adapt it by then. But he never turned up for his screen test, so... We were running out of time and had to make decisions based on the screen tests. The fact that he never turned up meant that he could never be considered.

This is a regret of mine. I think he would have made a better Heathcliff.

The screen test was really useful for Orla and Robert who gave a very good performance. They started dancing round the studio, shouting their heads off, screaming at each other and laughing out like mad. And we just sat there with our eyes wide open... But that never translated on to the screen.

I did think that they were a little bit old for the ages of the characters. In the book, they are teenagers, aren't they? The story is much more believable that way. Orla and Robert were both around thirty years old at the time. However, we thought we got the right cast then.

VH: Polly Hemingway, in the part of Nellie Dean, is fantastic in a number of scenes involving the children. I particularly liked young Heathcliff's bathing scene in the kitchen, at the beginning of the movie. But I also thought she was a bit old for interpreting the character of Nellie who is the same age as Hindley and therefore a teenager herself when Heathcliff first arrives at Wuthering Heights.

DS: There is a huge timescale of thirty years involved and this is a big problem. The budget was OK but not luxurious: it was a six-week shoot. We could not afford another actress to play the young, twenty-year old Nellie. Being in her fifties, Polly does not visibly age in our film. I think she did well.[371]

VH: Are you familiar with the 1970 version starring Timothy Dalton (Heathcliff), Anna Calder-Marshall (Cathy) and Judy Cornwell (as a young Nellie), where only the first part of the novel is adapted?

DS: I thought the 1970 version with Timothy Dalton was bizarre. For a start, it was not his first role. And he was not very young, I think, since he had already appeared in a television series entitled *Sat'day While Sunday* before.[372]

What I remember from then is a Heathcliff who was completely insane, ran around like a cave man and screamed like a cornered animal! I could not really relate to him.

VH: To me, the screen couple Timothy Dalton-Anna Calder-Marshall worked very, very well. The intensity of their feelings was devoid, I thought, of any excess of sentimentality. Anyway, I feel for good movies that have been destroyed in the press – and it was the case for this one!

DS: The film itself ended very inconclusively.

VH: It should have ended like the Luis Buñuel movie where Alejandro/Heathcliff, shot by Roberto/Hindley, dies in the crypt on Catalina/Cathy's tomb. And what became clear after interviewing Patrick Tilley[373] (the screenwriter) and Bob Fuest[374] (the director) is that their intentions had been completely thwarted. Instead of a dark ending where a heathen and demonic Heathcliff would have fallen dead from his horse's saddle into Cathy's warm grave, a happy ending (a Samuel Goldwyn ending) was imposed on them by the producers. Timothy and Anna – like the Laurence Olivier and Merle Oberon of the 1939 version – had to be last seen resurrected: a happy couple of ghosts walking on the moors towards their heaven.

DS: Obviously you have to bear in mind that it has been nearly ten years since I watched it, back in 1997, before shooting my own film. I cannot remember very clearly what the structure of this film (or of the Kosminsky film) were like.

I remember watching them and can only possibly recall what were my first impressions on watching them.

VH: Back then, what was your first impression of the Kosminsky version?

DS: I thought that Juliette Binoche was a bizarre piece of casting. I could not go past her as being completely miscast. I was aware of her rather clear French accent all the way through. She is wonderful in most things. She is just wrong for this role. I believe Ralph Fiennes as Heathcliff was not right either, being too much of the English stiff-upper-lip type to perform that role. I was not able to follow the story since I was forever wondering 'what is she (Juliette) or is he (Ralph) doing?' Anything that takes you away from the story and reminds you that you are watching a piece of artifice is an indication of failure. You start noticing the fancy tricks and you forget the story.

VH: To come back to your film, your decision to cast Crispin Bonham-Carter in the role of Edgar Linton paid off with a clever interpretation that makes him an interesting, multifaceted Edgar. He conveyed some deep emotion like Matthew and Sarah for their own characters.

DS: Crispin was excellent. I loved his Edgar. He was very proper, rather gentle. And again, it worked so much better in the second half. Crispin's death scene was very effective: it is almost as if he stopped being, as if the light went out of his eyes. It was a very good piece of acting. I also really liked him in the scene where Cathy was giving birth and died. The scene cut between Heathcliff in the orchard, thrashing at the branches, and Edgar in the library, sitting by the fire, listening to every sound of his wife's agonised labour. As the camera slowly came round on to Edgar's face and wrist, nothing more than a painful tingling ran on Crispin's face. His performance was terrific.

VH: The violent sequence of Cathy's delirium and Isabella's elopement works the same way. It cuts between Cathy, slowly losing her mind, at the Grange, in her marital bedroom, her bed strewn with feathers, and Isabella's dismal arrival at the Heights, at night-time, the gargoyles of the old house sneering at her. Then the scene cuts again between Doctor Kenneth cutting forcefully Cathy's hair to 'remove the fever from her brain' and Heathcliff raping Isabella.

  Was it all down to editing?

DS: This must have happened in the editing. I cannot remember exactly how it was in the script. I know there were a couple of scenes (one scene Isabella arrives, one scene Cathy gets her hair cut, another scene Isabella is raped), which were put together to increase the impact since I took part in the process.

  However, there were a lot of little motifs – some more successful than others – which linked people's destinies throughout the film. There was a particular moment in a scene that epitomised the predestined aspect of their lives. It did not work very well, but the idea was good. Young Cathy and young Heathcliff are running through the gorge leading to Penistone Crags. She, alone, walks towards the cave, hesitates, sees something: she is seeing old Heathcliff coming towards her. This motif was repeated at the end, when Heathcliff dies and is reunited with young Cathy at Penistone Crags, making it clearer that she had previously been meeting a ghost of the future.

VH: This special moment in this early scene where young Cathy 'meets a ghost of the future' also ties up with the other dream sequences,[375] which remain very personal. I liked the way you used Cathy's portrait as a door (or 'portal') between the present and the past, between the world of the living and the world of the dead.

When he dies, Cathy's portrait allows Heathcliff to leap into the past where he sees young Cathy (as you just said) at Penistone Crags. He can then bring her back with him, for all eternity, as the adult woman of the painting.

DS: The idea of their predetermined existences, which we tried to work into the story, did not come from me, but from the writer. Unfortunately, I do not think that these special moments, loaded with the idea of predestination, worked very well visually. A circular link tied in with the opening of the narrative when, just after Lockwood sees young Cathy at the window, Heathcliff is screaming 'You show yourself to him, why not to me?!' and with the ending, the day before Heathcliff dies, when, at long last, the ghost of Cathy starts appearing fugitively to him.

VH: I was sensitive to the presence of the child actors in your movie – little Heathcliff, Cathy, Hindley and, later on, Hareton.

How easy was it to find them and then direct them?

DS: The children were quite hard to find. The initial scenes with the children, we did right at the end of the shoot on the main location. It was a terrible week, everything slowed us up. And the little girl fell off a horse and got a concussion…

There were scenes that we shot which were just unusable and had to be thrown away – like, for instance, the fight in the barn where Hindley throws an iron weight at Heathcliff and knocks him under the feet of his colt. I did not really have the time to rehearse with them properly. But, somehow, we got away with it.

They were good, they were OK. When you look at them, she is beautiful but he just looks like a small, not fully formed Heathcliff! We had six weeks – two of which on location – and kept running out of time. You need time to work properly with children!

VH: Where did you find a location for the farmhouse, Wuthering Heights?

DS: In Yorkshire. I really wanted to find some place which was right up there, a house stuck upon the top of the moors. We went to look at the moor where the original story is set, but it is now strewn with wind farms. There is a house that can still be seen and which Anne and Emily Brontë were describing. Because of the wind farms, though, we could not use it.[376]

We ended up shooting in the Yorkshire Dales National Park, which is completely unspoilt. There are no pylons, for instance. It is like stepping back in time, two or three hundred years ago! After driving round there for about three days, we saw this ruined farmhouse, right on the top of the hill. We had to walk for half a mile across the moors to get to it.

Eventually, we built a road in order to be able to take our equipment up there. It was the house we used and it was, I think, a good choice. The actual interior of the house was derelict, having been empty for thirty or forty years. We had to patch the roof up, even though we only shot the exterior. We also touched up the barn and added a lot of details to the outside of the house itself.

VH: What about the little manor, The Grange? Was it in the same area?

DS: Yes, but we found it off the limits of the Yorkshire Dales National Park, at probably an hour's drive from the main location. We had the choice for this other house. The fact that it was a bit far away did not really matter. Since we had shot all we needed in the National Park, once finished, we then moved to this secondary location, The Grange, to do three or four days of exteriors.

We spent ten days, in total, on the main location – one full week for the exteriors of Wuthering Heights and a couple of days more for the few other bits, like the graveyard scenes. And it was two weeks on location followed by four weeks in the studio.

However, the library scenes at the Grange were not shot in the studio but in a big house, whose name I have forgotten, just North of London. We did not have the time or the money to mess about all that much!

VH: What was the budget?

DS: It was either 1.6 or 1.8 million pounds.

VH: Despite this average budget and this relatively short time in the Yorkshire Dales, did you get some inspiration from the moors?

What about Robert and Orla in their difficult parts of Heathcliff and Cathy?

Did they get some nourishment from the Dales?

DS: Yes, the moors played an inspirational part. One thing that struck with me is this whole idea of Heathcliff and Cathy carved as elemental figures. They are not real human beings. They are like a God of the Earth and a Goddess of the Air, who have been accidentally thrown into the human world. When directing the actors, I was quite keen on this idea that Heathcliff and Cathy existed in this primeval world.

Between takes, Robert and Orla liked to listen to music a lot. Orla used to listen to some piano pieces, just to get her head into that kind of fiery, crazy, elemental place. We went to a lot of trouble to find a very particular location like the cave of Penistone Crags and that entire little dale with the trees. However, I am still not convinced that we really made the most of it.

But didn't you mention the farm animals in the preliminary questions you sent me?

VH: Yes, I really liked the way you gave a realistic texture to the story by using the noises of a real farm and by playing on the (potentially) disturbing visual presence of the animals – as illustrated by the dead sheep that welcomes Lockwood when he first enters the farmyard of Wuthering Heights.

As in the 1970 version, you made palpable the harsh living conditions on a Yorkshire farm in the Georgian times.

And I found your aesthetical use of the animals reminiscent of Buñuel's imagery in his Mexican version of the novel (1953) – this is meant as a compliment! As in the Buñuel version, both your farm and your wild animals are endowed with some metaphorical meanings. I am thinking of the dead fox, impaled on the gate, which serves as an omen of Hindley's death (for Heathcliff and the audience). I am also thinking of the sow and its piglet, lying by a heap of rubbish. We see them through the saddened eyes of Nellie when she visits her former nursling, Hareton, after Heathcliff's return to the Heights.

DS: That is what I wanted, this realistic feel. I wanted mud, I wanted dirt and I wanted animals. The pig and piglet, both covered in dirt, is an obvious reference to the degraded state in which Hindley and his son are. I also wanted to convey the idea that death and decay had already set in around them. The presence of the dead animals conveys this idea. There is the dead fox, impaled on the gate, but there is also a line of dead crows hanging from the tree.

You mentioned the sheep but there is a little lamb standing next to the dead sheep – which, by the way, was not real – as Lockwood is coming up to the Heights. This exemplifies very well your idea of visual metaphor. Here is death but new life is struggling to escape it.

In actual fact, there is a lot of nurturing going on in the house at that point. Hareton gives Catherine a little puppy, which she is holding in her arms, just before Lockwood enters the Heights. I would have loved developing this theme of nurture and love further.

There were dogs earlier on as well when Lockwood turns up at the Grange. They were supposed to growl and attack him. But the dogs just kept licking him. It took an hour and a half to two hours to make them bark – or even behave in a remotely threatening way! The scene where Isabella's dog is left almost hanging belongs to the book but also fitted quite well with the new scenes that linked animals with places and characters.

VH: We have not spoken of Ian Shaw who gave a strong performance as Hindley.

DS: His dad, Robert Shaw,[377] was a prolific actor but also a well-known alcoholic, who allegedly drank himself to death. The scene where Hindley drinks himself to death stuck out to me as a very good piece of acting, which Ian modeled on watching his dad. He got into the idea of the drinker, drinking slowly to the point of being unconscious, but hanging on to consciousness so that he can still keep track of what he does.

VH: The stance which you took was that Heathcliff starts him drinking by serving him a tumbler of spirits and leaving the decanter in front of him, on the dinner table, after the death of his wife Frances.

DS: That was good, that scene worked very well actually. Each member of the family left the table one after the other. Heathcliff alone remained with Hindley who finally picks up the tumbler and starts drinking. I was pleased with the scene. It worked well visually and dramatically. The acting was very good too.

VH: How did the audience respond to your film when it was first broadcast on ITV and later on when it was released in VHS format? [378]

DS: On that evening, the viewing reports showed nearly ten millions of viewers[379] – which was a huge figure then and is even more so now. The video of the film is still selling in most video shops. As the film was partly financed by WGBH Boston, the video was released in America where it did well. In the UK, it was the pick of the day in about five newspapers. The reviews were OK: one or two articles were very positive, while another one (the scathing one) tended to say that the production was a bit like a school's play, being too literal and not particularly expressive of the film-maker's ideas.[380]

A year later, I harboured this idea of moving the whole story to Cuba where you have the poverty, the explosive characters and the music! But my agent said 'Move on!'.

At the moment, I have three different television projects, which are running quite well, and a feature film (a horror film) I will be doing this Summer. Movies have always been what I wanted to do.

Bowker-Giedroyc (2009, ITV1, 137 min)

As mentioned at the end of the Summary, 'the novel's more politically (and aesthetically) subversive components periodically ebb away', and they do exactly that in Peter Bowker's cultural translation of *Wuthering Heights*.[381] The drama was commissioned in 2006 by former Head of BBC Drama, Series and Serials, and newly appointed ITV Controller of Drama, Laura Mackie. In her interview with Kate McMahon, in February 2009, she explains:

> We've already started having conversations about how we can challenge writers to think about ways of making something that still feels like a real quality drama but perhaps avoids the costs [...] It is hard to keep doing singles as opposed to three or four-parters because it takes the same amount of work for one night only. We felt we weren't getting the return on them.[382]

This 2009 version was honed and executive-produced by Peter Bowker, the much sought-after television screenwriter of the early to late 2000s. Even though he more recently tackled the script of the BBC One political drama, *Louder than Bombs* (2014), based on the 1996 Manchester bombing, Bowker had originally walked in the footsteps of television veteran Andrew Davies. He would deliberately divide his time between the adaptation of British classics – from Dickens' *A Christmas Carol* (2000) and Chaucer's *The Miller's Tale* (2003) to Shakespeare's *A Midsummer Night's Dream* (2005) – and the creation of original screenplays such as the touching drama, *Flesh and Blood* (2002), the musical thriller, *Blackpool* (2004), and the Iraq war-inspired, *Occupation* (2009). His adaptation of *Wuthering Heights* in 2008-2009 consisted in a stand-alone television movie, advertised by ITV as a two-part serial and broadcast on the last holiday weekend in August 2009 (Sunday 30th and Monday 31st). It would gather, between 9 and 10.30 pm, a fifth of all the audience ratings (approximately 4 million viewers on each night). This was equating the numbers that the 1998 version by Skynner had reached in only one single night. As in his (truly) serialised dramas shown earlier that Summer, the three-part *Occupation* (June) – and the six-part *Desperate Romantics* (July-August) based on Franny Moyle's *Desperate Romantics: The Private Lives of the Pre-Raphaelites* – he foregrounded the Profane and Intimate as opposed to the Sacred in what became Heathcliff's 'fatal love affair'. While taking Emily Brontë's love journey at the extreme top of the Gothic architecture of *Mise en Abîme*, he distanced himself from the ontophanic domain of myths, and came closer to the 1939 Olivier movie. Bowker's answers to journalist and television critic Stuart Jeffries in the *Guardian* article, 'Sex and Rebellion: *Desperate Romantics* Writer Peter Bowker on His New BBC Drama' (2009), reveal his intimate approach to the character of Heathcliff, and his overwhelming desire to please the television medium:

> Why did *Occupation* work? 'Because, and I shouldn't tell you this, it isn't a war drama at all,' Bowker admits. […] 'It's not really about war at all. It's about three men in the aftermath of the Iraq war and why they have to go back there. One goes for love, one out of conscience and the third for money.' […] 'I loved Heathcliff's class hatred. I really wanted him to stick it to those toff bastards. But of course that's a monstrous misreading. To revenge yourself for the crimes of others against you on innocent children is immoral. But I still want Heathcliff to get under your skin to such an extent that what he does makes sense to you, and you can go with it some of the way, at least.' […] 'My partner Kate, who's a couples [sic] counsellor, gave me the key to the novel's psychology. She said that the wonderful speech in which Cathy says 'I am Heathcliff' tells you the relationship isn't going to work, even though it's one of the great declarations of romantic passion.' […][383]

While Bowker privileges the Intimate (and consensual) over the Sacred (and contentious), the Mythical Components retreat into the hyper-romantic book that accompanies Catherine from the Grange to the Heights, Walter Scott's *Ivanhoe*. In this translation of *Wuthering Heights* into a television drama, the sociopolitical backdrop is replaced by a Christian rationale upheld by Old Earnshaw, the parish benefactor and Heathcliff's Saviour. The latter simply cannot comprehend his son Hindley's selfish rejection of the gypsy orphan, and there is no Mrs Earnshaw to give a dissentient moral counterpoint to her husband's all too caricaturally virtuous stance.

The *Mise en Abîme* architecture, in the absence of Lockwood (or his substitutes, all the other intra-diegetic observers), is flattened into a two-dimensional storyline where the ill-treated young adults of the second generation, Catherine (Rebecca Night) and Linton (Tom Payne), are given chronological pre-eminence and screentime advantage over the children of the first generation, Cathy (Alexandra Pearson) and Heathcliff (Declan Wheeldon). The latter's performance is confined to four sequences: the young Heathcliff's arrival at Wuthering Heights, his introduction to the small-minded world of the parish church, his beating by Hindley in the farm's stables and his subsequent involvement in a fight with a village boy, which leads to Earnshaw's decision of uprooting the evil influence of Hindley from the farm. Bowker's narrative choices are problematic since they alter the dynamic structuring of the source novel and of its unconscious. Rather than establishing the children's connection with the moor, Bowker dictates their belonging to the microcosm of the parish. Therefore, in a short cut that does not leave much room for character development, his narrative choices not only validate the legitimacy of Heathcliff over Hareton as future

Master of the Heights, but also that of the young established lovers (who are, in both generations, the highlighted characters) over the children. This entails that the very notion of *Trespassing* is jettisoned while *l'amour sensuel*, which comes to subsume the relationship of the emblematic lovers, Heathcliff (Tom Hardy) and Cathy (Charlotte Riley), does not carry at all with it the notion of *Interdict*. The Third Bataillan Theme (BT), which is also associated with 'a *taboo on games of sensuality*', is voided of its substance:

> In order to give a better representation of Good and Evil I shall return to the fundamental theme of *Wuthering Heights*, to childhood, when the love between Catherine and Heathcliff originated. The two children spent their time racing wildly on the heath. They abandoned themselves, *untrammelled by any restraint or convention other than a taboo on games of sensuality*. But, in their innocence, they placed their indestructible love for one another on another level, and indeed perhaps this love can be reduced to the refusal to give up an infantile freedom which had not been amended by the laws of society or of conventional politeness.[384] [Emphasis added]

In Bowker's dramatised version, the neutralisation of Heathcliff's pain and anger, which is necessary for the final reunion of the second generation of lovers, also relies on the aptitude of the couple, Hareton (Andrew Hawley) and Catherine (Rebecca Night), to provide an ordinary and dispassionate counterpoint to the Gothic love story that has transformed Heathcliff into a Monster. Although the acting is convincing, and even compelling in the dénouement scenes involving Heathcliff, Catherine and Hareton, both Hardy and Night represent a challenge to the BBC television viewers as actors-impersonators, not in terms of their performances *per se*, but in terms of *performance authenticity*.[385]

Tom Hardy brings into his incarnation of Heathcliff 'the intertextuality of his casting' as the Bill Sikes of Andrew Davies' *Oliver Twist*, a mini-series broadcasted during the Christmas season 2007 while Rebecca Night's fresh impersonation of Fanny Hill, in Coky Giedroyc's hugely popular two-part eponymic serial of the Autumn 2007, gives her Catherine a share in the cursed *amour sensuel* that actually ruins Cathy/Riley prospects with Heathcliff/Hardy.[386] Hardy and Night's consecutive lack in *performance authenticity* is redeemed by Giedroyc's *mise en scène*, and by the quality of the photography and camerawork. In the final sequence, as Catherine/Night slowly checks the now emptied family sitting-room, and Hareton/Hawley sees to the horse-driven carts loaded with trunks and furniture, Ulf Brantås' camera, which originally introduced the viewers to the coarse grasses of the heath and showed the way to Cathy's upstairs bedroom at the Heights, prefigures the hyper-subjective camera style that would be used, two years later, by Robbie Ryan for Andrea Arnold's *Wuthering Heights*. In a finale reminiscent of the last shots of Alejandro Amenábar's *The Others* (2001), the righteous couple, Catherine/Night and Hareton/Hawley, relinquishes Wuthering Heights to their unruly inhabitants, Cathy/Riley and Heathcliff/Hardy, who have long lost their ghoulish edge.

Chapter 25

The *Anti-Period Dramas* in Britain and Abroad: *Displacement* and Compensation

Wainwright- Sheppard (2002, BBC One, UK, 3x60 min)

The concept of the three-part serial, *Sparkhouse*, originated in screenwriter Sally Wainwright, who convinced Nicola Schindler, executive producer for Red Productions, then Gareth Neame, Head of Independent Commissioning at the BBC, that 'effecting a reversal of sex' between Heathcliff and Cathy would lead to the creation of an original contemporary tale, with an unforgettable woman character at its centre.[387] The screenplay revolved around the character of Carol/Heathcliff, a role tailor-made by Wainwright, for Sarah Smart. Impersonating Virginia Braithwaite, in *At Home with the Braithwaites*, Smart had demonstrated her feisty temperament to the screenwriter, as well as thoroughly impressed director Robin Sheppard.[388] The young actress could also capitalise on her interpretation, in the 1998 period drama directed by David Skynner, of a very touching yet strong Catherine, reminiscent of the unsaccharine character developed by Cathryn Harrison and Peter Hammond for the 1978 *Classic Serial* production.

Because of the contemporary, *anti-period drama* nature of her script, Wainwright was intent on composing a socio-economical backdrop suitable for a genuine adaptation of *Wuthering Heights* on to the small screen. It was an 'epic love story about two people, Carol Bolton and Andrew Lawton (Joe McFadden), who come from contrasting classes and cultures', an epic love story that, above all, had to be credible.[389] Wainwright placed on the foreground Paul Lawton's medical practice and his wife's primary school, the village's public house and secondary school, and the shops and registry office of the nearby town of Halifax, while leaving in a fuzzy background the campus of Manchester University and the Council towers of Bradford (or is it Leeds?). Wainwright would translate more than adequately the First Mythical Component (MC1) by establishing the contrasted topographical landmarks of the Boltons' struggling cattle farm and the Lawtons' upper-middle-class villa, as well as by inventing the 'Den', a derelict house on the moors that stretch between Todmorden, Hebden Bridge and Saddleworth. The 'Den' was to re-create, for Carol and Andrew, the Penistone Crags of Cathy and Heathcliff.

Wainwright's *Wuthering Heights* tale of class prejudice, child abuse and individual resilience did not require 'representational' authenticity only to be exempt from melodrama, a tonality that the contemporary *anti-period drama* does not really allow. It also required 'performance' authenticity that was delivered by Smart, in the lead role, and a team of excellent actors in the midst of which stood out Celia Imrie (as Mrs Lawton, Andrew's mother), Richard Armitage (as John Sandring, Carol's husband-to-be) and Holly Grainger (as Lisa, Carol's daughter), the future Diana Rivers of Cary Fukunaga's *Jane Eyre* (2011). Additionally, the sequences where Carol and Andrew are reading poems written by Emily Brontë or even passages from *Wuthering Heights* in their Den, this liminal place in the middle of the moor associated with their childhood, actualise very effectively the Second MC (MC2), and reproduce symbolically the three-dimensional Gothic Figure of *Mise en Abîme* of the novel. Wainwright's reluctance to acknowledge that the dynamic structures of *Wuthering Heights* did more than appear accidentally in her script seems to mirror the limits of the public's acceptance, in the context of a British production, for what we have defined and classified as the Mythical Components (MCs) and Bataillan Themes (BTs). Her condensed three-part serial is ambitiously depicting the complex character of Carol, who is more an epitome of the two Cathys than the feminised Heathcliff initially envisaged, and retracing her Initiatory Love Journey (MC3). It is also following the multiple learning trajectories of the intra-diegetic outsiders' substitutes, whose qualities, in the absence of Lockwood and Ellen, have been distributed between Andrew's dad, Paul (Nicholas Farrell), Andrew's young wife, Becky (Camilla Power), and Carol's teenage daughter, Lisa (the excellent Holly Grainger).

The abundance of good ideas in the 2002 serial of *Wuthering Heights* testifies to Wainwright's natural translative skills that she applies, conscientiously, to the source text and its unconscious (or its dynamic structures). She even operates a transfer of personalities between the Hareton of Brontë's novel and the solid farm worker, John Sandring, who becomes Carol's husband and allows her to buy the farm back, after the death of her oppressive and incestuous father. The three episodes of this 180-minute long version, packed with action and emotion, were, nonetheless, heavy to digest over a single weekend of broadcast on BBC One. Further, the poor soundtrack of the opening and closing credits could not have been conducive to keeping the indecisive spectators on the channel. According to the BARB figures, and the blunt review by John Plunkett published in *The Guardian* immediately after the serial's broadcast, *Sparkhouse* proved to be 'a damp squib' with 'an audience of just 4.4 million – fewer than one in five viewers' but slightly better than the 2009 ITV *Wuthering Heights* written by Peter Bowker.[390] Wainwright's British contemporary *anti-period drama*, out of an excess of fervour for the unconscious of Emily Brontë's novel, might well have exhausted the less adventurous viewers amongst its potential audience. However, with an original strategy of compensation by **displacement** that de-multiplies the Third Mythical Component (MC3) into several self-revelatory journeys, and draws on all three BTs, she also boldly stated that, in the words of George Steiner, 'the source-text possessed potentialities, elemental reserves as yet unrealised by itself'.[391]

Sally Wainwright (*Sparkhouse*, 2002)

Written answer, 30th Oct, 2002

VH: The bold and strong idea of the three-part serial, *Sparkhouse*, came from you. In the 'press pack' made available by the BBC on their website,[392] Nicola Schindler (the executive producer from Red Productions) explains that it was an idea which you had always wanted to develop and turn into a screenplay. You gave her to imagine a contemporary tale inspired by *Wuthering Heights* and you won her over.

She sees it as an 'epic love story about two people who come from contrasting classes and cultures'. She also explains that the parallels with *Wuthering Heights* 'are not exact in any way' but that the Yorkshire Moors which you know so well are as present and essential as they are in the Georgian period novel.

How did you find the title of your serial?

SW: Near where I grew up, there was a road up on the moors called 'Sparkhouse Lane' and I always thought what a fantastic name it was.

VH: The next step for you and Red Productions was to find a broadcaster. Apparently, it was not difficult to convince Gareth Neame (BBC's Head of Independent Commissioning) to commission this original and modern television drama – which also drew on some perennial themes and was inspired by a literary classic.

You initiated the television project of *Sparkhouse* and, as a co-producer, were able to remain at the heart of the production process until the end. Isn't it quite exceptional for an author to enjoy such creative freedom, especially when it comes to the adaptation of such a text?

How did you influence the choice of the locations (Todmorden, Hebden Bridge and Saddleworth Moor) in West Yorkshire?

SW: Most TV dramas come from original ideas. It's not exceptional.[393] Where I am lucky, I think, is that I am able to write about subjects that are not to do with the usual TV diet of doctors, lawyers and cops.

I didn't choose the locations. The location manager chose the locations. That was where it was set, so it made sense to film it there.[394]

The landscape was important. However, for me, the biggest thrill (regarding *Wuthering Heights*) was trying to create a female character who was as wild and angry and screwed up, but also as magnificent and passionately in love as Heathcliff.[395]

VH: Your screenplay revolves around the character of Carol Bolton/Heathcliff and you tailor-made Carol for the actress Sarah Smart – who demonstrated her feisty temperament in the part of Virginia Braithwaite (another of your creations) in *At Home with the Braithwaites*.[396]

Having Sarah in the lead role was a necessity. What about the rest of the actors? Did you have a say in their casting? Did you play any part in the direction and *mise en scène*?

Your former teacher, the painter Peter Brook, came to visit you (and the rest of the team) when you were on the moors, above Hebden Bridge and Todmorden, trying to find the right buildings[397] for the farm and the 'Den' – the Penistone Crags of Carol and Andrew.

How was his exhibition, 'Peter Brook in the Pennines', at the Smith Art Gallery in Brighouse, influenced by *Sparkhouse*?

SW: I was at all the auditions, attended all the rehearsals and was out on location as often as I could be. I have a very good working relationship with Robin Sheppard.

Peter came out on location a couple of times. I think he liked the story and he liked the fact that it was set in his part of the world. I suppose you'd have to ask him how it influenced him.

VH: You staged a couple of scenes where Andrew and Carol are reading aloud some key emotional passages of *Wuthering Heights*. Andrew also reads for his secondary school pupils and for his newborn baby.

Can you remember what he reads to baby Tom?

Was *Sparkhouse* intended to draw attention to the source novel again and appeal to the younger generations?

SW: It is by Emily Brontë. As far as I recall it is a poem simply entitled 'Stanzas'. But I didn't really think about the audience with *Sparkhouse*. Hopefully if you're telling a good story it should be of interest to lots of different types of people.

VH: The incest motif and the questionings about Heathcliff's origins are constantly brewing under the surface of the novel. Furthermore, in the next generation, Catherine Linton weds her cousin Linton then falls in love with her other cousin, Hareton.

In his script (1970), Patrick Tilley chose to bring the motif to the surface. At the start of the movie, there is the broad hint that the child Heathcliff could well be the result of 'Mr Earnshaw's own doings' then, later on, that Cathy's baby might be the child of Heathcliff – on Heathcliff's coming back

266

to the Heights with Isabella, Joseph says in an aside to Heathcliff that 'Edgar is waiting to see the colour of its eyes'.

In *Sparkhouse*, the incestuous abuse of Carol by her father, Richard, when she was a very young teenager is not merely an emotional hand grenade. It is clearly a dramatic element ('the dirty family secrets' of Sparkhouse farm) which is key to your own plot.

Would you like to comment on that?

SW: I hadn't even thought about incest in *Wuthering Heights*.

VH: What did you think of the fast beat of the soundtrack which accompanies the last sequence of each episode and matches the pace of the end credits while they quickly scroll up the screen?

SW: I thought the music was disappointing. But we were having to make the best of a bad lot as the soundtrack could not be changed. We were on the verge of appointing a different composer when the BBC sprung a very sudden transmission date on us, so we had to stick with him. That was very unfortunate.

VH: In the last episode, Carol's daughter, Lisa, and Andrew's wife, Becky, have a go at interior decoration as Lisa wants to make a nice surprise to her mum who now owns Sparkhouse farm: they are painting one of the walls red.[398] What triggered this idea?

SW: I thought it was funny because the show was made by Red Productions. It made me smile when Carol saw the wall and said 'It's very red' – meaning the whole show was because it was made by *Red*. It was my private tribute to Nicola Schindler.

Rivette-Schiffman-Bonitzer (1985, France, 126 min)

In the Autumn of 1985, the cinema theatres were empty on *Hurlevent*'s release in France. The film went practically unnoticed there as it would later go unnoticed abroad. At the time, *Hurlevent* did not only fail to bring Jacques Rivette much recognition (or much of an audience) since it was above all, for the initiator of the project and *metteur en scène*, a creative and personal hurdle. He had lost his close friend François Truffaut while shooting the film, and would have to wait until *La bande des quatre* (aka *The Gang of Four*), in 1988, to meet with an artistic and commercial success again. In the interview that Rivette gave me in December 2002, the French cinéaste recalled that he first read Emily Brontë's novel in its original French translation by Frédéric Delebecque, *Les Hauts de Hurlevent*, as he was a eighteen- or nineteen-year-old, around 1946.[399] Four decades later, inspired by the celebrated series of Persian-ink drawings that the painter Balthus had created in the early 1930s when sharing a flat with Pierre Leyris (who would later become the other major translator of *Wuthering Heights* into French with his *Hurlemont des Vents*), Rivette was finally ready to direct his own cultural translation of Brontë's novel, *Hurlevent*. The 126-minute film shifted the Yorkshire *moors* of the First Mythical Component (MC1) to the backdrop of the Cévennes, an area strongly associated, in France, with a stern rural Protestantism and characterised by a wild, sun-drenched landscape, the *garrigue*, where isolated farms can be several miles apart.

From the inception of the project, Rivette had only been interested in adapting the first chapters, until the death of Catherine in Vol. II, Chap. II, and started working on the script accordingly with his habitual co-scenarists Pascal Bonitzer and Suzanne Schiffman. He probably also asked Bonitzer to remove the source text's most prominent intra-diegetic outsider, Lockwood, and the character of Ellen/Hélène, following a translative logic of **compensation** similar to Tilley's, progressed in the leading role of intra-diegetic outsider, drawing the film spectators into the story even though she never interprets it for them, and remains quite enigmatic herself. Rivette's *mise en scène* harmonised the minimalism of the sets and locations – consisting of the Séveniers' Cévenole farm (La Croix des Grives), the Lindons' country house complete with tennis court and the wild garrigue that stretches in the middle – with the simplicity of the actors' performances. Synchronised with the first half of 1930s, that period that witnessed the last meaningful Surrealist works in France, the actors' play seems to be 're-enacting' the tragic events of Emily Brontë's tale, and continuing the revelatory experience of the English poets of the *Lyrical Ballads*.[400]

To regain the three-dimensional feel of the novel's *Mise en Abîme* architecture, which the heightened presence of Hélène/Sandra Montaigu could not bring back on its own, Rivette and his co-scenarists introduced three dream sequences, one at the beginning that stages the spying of Guillaume/Hindley on the sun-bathing lovers, one in the middle and one at the end of the picture that uses the motif of the broken window and of the hand. The middle oneiric sequence is the most impressive, and was discussed with Rivette at the time of the interview. It marks a natural elliptical turn in the action – the three-year separation of Roch/Heathcliff from Catherine/Cathy – and actualises intensely the Second MC (MC2). The blurring of the frontiers between conscious and unconscious perception is, in this breath-taking sequence, made palpable by a visualisation of the physical state of passion, which gives a poignant insight into Catherine's and Roch's psyches, without any need for words. The youth of the actors was essential to Rivette for the credibility of *Hurlevent*, his cultural translation. To him, performance authenticity meant young actors, not only for the main couple, Catherine and Roch, but also for the parallel couples, Olivier and Isabelle, and Hélène and Guillaume, since his (and his co-scenarists') critical reading of *Wuthering Heights* corroborates the readings of Charles Sanger and Patrick Tilley.[401] This led Rivette to adjust the age and status of the housekeeper (and family friend), Hélène, to whom Sandra Montaigu lent some empathic discernment and confident attractiveness.[402] The emphasis on the strong individuality of Ellen's character, which is, however, embedded in the narrative of *Wuthering Heights*, activates, on screen, the dynamic of a latent love interest between Hélène and Guillaume. By inscribing the attachment of Roch and Catherine (and of Olivier and Isabelle) in a pre-existing pattern, the unfulfillability of intense childhood relationships in adulthood becomes part of the destinies of the inhabitants of the garrigue, where survival and reproduction is a constant struggle. Therefore, the Bataillan Themes (BTs) all translate very well despite the conciseness of the second part of the film, which revolves around Catherine's new life with her very young husband and sister-in-law, and the condensed oneiric ending applied to the love journey(s). Still, this short-handedness, which reflects partly the difficulties that Rivette encountered to finance and finish the film, impacted on the archetypal structures of the novel – in particular, on the Second and Third MCs (MC2 and MC3).

MC2, boosted by the three structural dream sequences, is considerably altered by the staging of grown-up characters only, even in the dream and festive scenes at La Croix des Grives, and even more so in the Lindons' household, which is depicted as a place a little more immersed in societal exchanges than the farm of the Séveniers. This invisibility of the children constitutes a translative flaw in the story-retelling, which is most noticeable in the sequence of the 14th July celebrations where a cast reduced to its bare minimum cannot re-create the bustle and vibrancy of Christmas Day 1777 in Emily Brontë's Chap. VII, Vol. I. Further, despite Hélène's romantic journey that duplicates Catherine's love journey, MC3 is weakened. Owing to its own dramatic under-development, Hélène's parallel journey can only *compensate* to some limited extent for Catherine's and Isabelle's truncated stories. Rivette, aware of the melodramatic weight of some of Cathy's emblematic lines (such as 'I *am* Heathcliff', in Vol. I, Chap. IX, or 'I wish I could hold you till we were both dead', in Vol. II, Chap. I), believed in allowing Hélène to be in a sisterly relationship with Catherine and to have experienced, in a quieter mode, some similar feelings. He also believed in the simplicity of the *mise en scène* to alleviate the most tragic scenes; Hélène is ironing the linen when Catherine declares 'I *am* Heathcliff', and removing cakes and tarts from their baking tins, when Roch comes to confide in her before the 14th July party starts.

In this French *anti-period drama* where the cinéaste is at liberty to choose settings, period and unsophisticated costumes, independently of the source text, Rivette succeeded in showcasing the youthful grace of the characters against a neat backdrop of garrigues and minimalist farm sets, and managed to reveal the depth of their inner tragedies, as compellingly as Balthus had done before him with his unembellished illustrations of *Wuthering Heights*. A concession to lyricism, nevertheless, can be found in the use, in the soundtrack, of the magical accents of *Le mystère des voix bulgares*, the album of a Bulgarian choir reminiscent of Béla Bartók's ethnical pieces. The voices of the choir echo at some epiphanic moments of physical freedom and spiritual release: Roch and

Catherine, dashing across the garrigue on their way to the Lindons' house, or Roch putting his entire arm through the broken pane of Catherine's old bedroom window, as he is about to be reunited with her, at the end of the film.

Jacques Rivette (*Hurlevent*, 1985)

Jacques Rivette's Adaptation of *Wuthering Heights*

This interview was published in Senses of Cinema in the Autumn of 2003.

On a crisp and sunny Monday, just before Christmas Eve last year, I met Jacques Rivette in Paris. I sat with my idol for an hour-and-a-half interview in a typical Parisian café, Le Café des Ateliers, situated near Boulevard de la Bastille (in the 12ᵗʰ district) and buzzing with the sound of animated conversations.

I was so determined to discuss his mid-1980s movie, Hurlevent, which inspired me to start my university project – 'The Cultural Translations of Wuthering Heights' – that I had begun to correspond with him two months earlier. Rivette remained terribly silent, although he did read my mail. As I learnt later, he is rather hesitant when it comes to responding in writing – and a less demanding electronic correspondence is simply out of the question: 'I am too old' he would complain.

Nevertheless, once I finally managed to take advantage of a golden opportunity to talk with him on the phone, being a nice guy (generous with his time and fairly open to new people) he acknowledged my two letters and instantly granted me the interview I was hoping for… and 'instantly' means 'instantly': I rang him at 2 pm and interviewed him shortly after three o'clock on the same day!

VH: The main topic I would like to explore with you concerns the episodes of the novel that are difficult to translate into images, and their oneiric equivalents in the first, middle and last scenes of *Hurlevent*. According to what Pascal Bonitzer said in a filmed interview,[403] the first scene was inspired by Bataille. While leafing through the critical publications about Emily Brontë's novel, he came across a book by Georges Bataille, *The Tears of Eros*…

JR: Well, it is not a book but an article, a very long article, which he collected later in a book called – I think – *Literature and Evil* (*La littérature et le mal*). Pascal knows Bataille far better than I do, since he was the subject of his Ph.D. thesis. As for me, a long time ago, I merely skimmed through this article by Bataille. The truth is that I am not very familiar with it.

VH: So there was this article, but also an illustration, apparently a reproduction of a painting by Poussin…

JR: No, it was not exactly an illustration. It all started – and it has been stated in a number of interviews, so you must have heard about it – it all started when I had no plans to shoot an adaptation of *Wuthering Heights*, or anything else, for that matter. It was after…

VH: …the Balthus exhibition?

JR: It was after *Love on the Ground* (1984). I had just finished the editing – it was probably at the end of 1983 or at the start of 1984. I believe it was at the start of 1984 that the Balthus exhibition took place in Beaubourg.

So I went to this exhibition. Seeing as he's a bit of an eccentric and all that, I am very fond of Balthus. So I went to the exhibition which was actually superb. I already knew the drawings produced by Balthus for the book that the *Gallimard* editions had intended to publish at the beginning of the 1930s – around 1932 or 1933, I think. These drawings, by the way, were more or less contemporary with Buñuel's first desire to film the novel… I believe he had already written the screenplay…

VH: Which he only shot 20 years later…

JR: Yes, but still, his screenplay was written at the time in question. So it was in the air for this little group, and Buñuel, Balthus and so on knew each other. They used to gravitate around the Surrealists, while retaining their independence. And then, although I had already seen some reproductions of the drawings, the Balthus exhibition of Beaubourg featured a small, separate room – a kind of 'tablier', as one says in old French – where they actually displayed all the Balthus originals – the ink as well as the pencil ones, the final drawings as well as the sketches.

And I was struck by the fact that Balthus enormously simplified the costumes and stripped away the imagery trappings which are so much foregrounded in the Wyler movie. I wondered why nobody had ever made a movie in which Catherine and Heathcliff were the age they actually are in the novel. Because in the Wyler movie they are 30 and in the Buñuel movie 30 or 40.

Therefore they are adults, and it does not mean anything. Well, it does mean something, but something completely different. So I felt like making a movie with some very young actors. I started with this idea in mind and made the first adaptation – well, maybe not the first one because there are adaptations that I have never seen – in which they are their age.

It was a novel that I had read, like everyone else, when I was 18 or 19 in its classic French translation, the famous translation called *Les Hauts de Hurle-*

Vent by Frédéric Delebecque – which is quite a good translation. It is a free translation, a kind of adaptation for the French language, which, as far as I know, is pretty faithful. The only criticism that I might pass, very quickly, on the translation of Monsieur Delebecque is that everybody uses the 'vous' form while, theoretically, between Catherine and Heathcliff...

VH: Maybe because there is never any 'thou' in the novel...

JR: Of course. Still, I don't know which Emily Brontë would have chosen. Because, on the one hand, when she writes the novel in 1840 or so, the 'you' is very strong above all, maybe, in a Protestant environment. On the other hand, I really find that the 'tu' form comes more naturally. In English, I am not sure...

The important thing is that it was quite a good translation. In fact, I had gone through it and tried to establish a few comparisons with the English text. And of course, when we prepared the movie, I bought the English text and compared. But then, I deliberately decided not to re-read it.

So I started with this idea in mind, and talked first to my producer, Martine Marignac, with whom I had already made *North Bridge* (1980) and *Love on the Ground*, and then to Pascal and Suzanne Schiffman, with whom I had worked on *Love on the Ground* too. (*Love on the Ground* was the first movie I had made with Pascal; Suzanne, I had known for years and years.)

But I had decided not to re-read it: I asked Pascal to summarise it for me. I only wanted to have the outline of the story and of the characters, that's all. And from the start, I told him: 'Only the first part', because I knew about the second part. I had a very strong memory of the Wyler movie – because I hate it – and of the Buñuel movie because, as you know, I find it very beautiful. The characters are 40, but still, the movie remains very, very powerful.

VH: After *Love on the Ground*, which was shot in the suburbs of Paris, did you feel like heading southwards for *Hurlevent*?

JR: Indeed, I was tempted to make a film with a very stormy atmosphere, with the idea of 'wilderness' which had been completely absent from all I had done before.

VH: Did you know this impressive stone building, this farm or 'mas', prior to the shooting?

JR: No, not at all. In fact, Suzanne and I spent a long time looking for it. In the end, we eventually found this place. It was mainly the farm that mattered: the farm – as you probably know – is in Ardèche whereas the mansion, which should be nearby, is actually located 100 kilometres below, between Nîmes and Montpellier, near Sommières. However, it became obvious at an early stage that we would not find the mansion in Ardèche, or the setting for the farm near Sommières. I personally like the farm a lot. The mansion was trickier. It was a very, very tough shoot; it was a terrible shoot.

VH: That is why I was unwilling to bring up the subject – I wonder if the novel can cast a bad spell. Why were there so many difficulties? You said that between the actors and the technicians…

JR: We experienced difficulties at every level. The first difficulty regarded the financing of the movie. We had been denied the 'avance sur recettes' (advance on takings[404]) – which rarely happens – and that was it. The movie got eventually financed thanks to the intervention of Jack Lang [then Secretary of Culture], who organised a donation directly from the Ministry and also thanks to Claude Berri who entered into co-production. The latter told Martine, 'Well, since you have trouble finding the money, don't worry I'll co-produce the movie with you,' and brought along the distributor, which was crucial. So that was the first difficulty. The second one was quite typical; it consisted of finding the young actors. In this respect, I was happy, it went well. The only problem was that, in the end, there were, as you know, these two actresses whom I hesitated between for Catherine's part. There were Fabienne, whom I finally chose – and Emmanuelle Béart. It is unlike me but, for that movie, I did quite a few video tests with a lot of young actresses and young actors. Funnily enough, the only one that we did not see at the time was Binoche, and who knows, if we had met her… She played in *Rendez-vous* (Andre Téchiné, 1986) almost immediately after our shoot ended. It was the same director of photography.

    It was pretty clear to me that our Heathcliff would be Lucas, because he had this sort of 'peasant' demeanour when he walked, when he talked, in fact, in his whole being. To me, he was a kind of emblematic Heathcliff, even if he did not possess at all the dark romanticism of Emily Brontë's Heathcliff. And if I eventually chose Fabienne rather than Emmanuelle – they were both smashing during the tests – it is only because I found that there were more affinities between the two…

VH: Between Lucas and Fabienne?

JR: Between Lucas and Fabienne, rather than between Lucas and Emmanuelle. First and foremost, the fact that Emmanuelle was a brunette whereas Lucas and Fabienne were both fair. That, I believe, tilted the balance in favour of Fabienne.

However, the shoot was very tough. Since the financing of the movie had taken so long, the crew was everything but homogeneous. The project did not start rolling till very late and the director of photography I wanted was not available. So, at that stage, we hired the people who happened to be free. And it did not go well with Berta: he is a very good director of photography but failed to breathe team spirit into the crew. And the young actors felt traumatised because they could sense that something was not working, that it was a shoot ridden with unnecessary crises.

On top of that, the thing which, of course, weighed considerably on the shoot, was that Suzanne – me too, but even more so Suzanne, since she was closer to him – knew that François Truffaut was dying. And he died a few days after the shoot ended. So, for the whole length of the shoot, every single day, we were expecting to receive the phone call that would tell us 'François has died…' it was a truly harrowing situation.

Therefore, everything conspired to make it a difficult shoot. And there were scenes that were not shot because Monsieur Berta, who is a great director of photography, also needs loads of time. So we either simplified or deleted. In my opinion, some important scenes are missing towards the end, in particular when it comes to Catherine's long illness, which is really too elliptical – and it is so powerful in the novel. That is why I believe that we did a better job of the scenes which were shot on the farm than of those which were shot in the mansion where things went even worse with Monsieur Berta and the heat of the spotlights.

I was personally exhausted and, towards the end, as if it were not enough, a cold epidemic broke out, which means that we were all half-sick. Fabienne was very sick, so sick that it could well have helped her to play the dying Catherine…

VH: But how long did the shoot take? How much time did you have?

JR: I forget. It was probably no more than a six-to seven-week shoot. We had so little money, so little!

VH: And who was your distributor?

JR: It was the distributor who, at the time, depended on Claude Berri, that is to say the *AMLF*. So the distribution came with Claude Berri. But the film was not a success and, after the editing – it was painstakingly edited – we also faced serious difficulties with the sound mixing. The sound mixing was a nightmare. As it is the sound after mixing was pretty bad – it was already average on shooting. Well, it did not compare with what I usually got from the other sound engineers. But the worst came on mixing: we mixed in a bad auditorium and what we heard did not synchronise at all with what we had on the film. We then had to digitalise and reinforce the direct sound… Well, for me, it is too late now to give it another try! I'd better get on with it…

Therefore, it was really a problematic movie. Actually, I don't know what to think of it. I haven't seen it since and would be very much afraid of seeing it now. I believe there are things that I truly like and others that are definitely not so good. But then I hope that we did capture the force of the subject and of the novel, since we remained faithful to it and also since the actors were good. I really like the actor [Olivier Cruveiller] who plays the older brother and with whom, incidentally, I worked again in *Jeanne*.

VH: How did you choose Olivier and Isabelle, since there is an opposition between the fair hair of the Séveniers (Catherine, Heathcliff) and the dark hair of the Lindons (Olivier, Isabelle)?

JR: I must confess I did not pay attention to that…

VH: Those actors are also very, very young.

JR: Yes indeed, since we had met Olivier by chance, during a casting for Heathcliff. But I told him from the start that he must play the elder brother…

VH: Hindley…

JR: Well yes, but he had to be given a French name. I called him 'Guillaume', if I remember correctly. He is an actor whom I like a lot and chose again for Jeanne. Also, the actress who played the maid-servant [Sandra Montaigu] had a small role in *Love on the Ground*. And I really like the actress who played Isabelle, too [Alice de Poncheville]: she was only fifteen when we made the movie. She had never done anything. And I think she had the potential, but it just did not happen. Later on, she made one or two movies in which she was good too, but she did not persevere. She lacked the vocation. It was fun for her but she lacked the vocation.

VH: And the costumes were quite remarkable, weren't they?

JR: Yes, I agree. But this had a lot to do with Lydie Mahias, our script-girl, who, in actual fact, did far more than one job. I had decided pretty soon that it would take place in the 1930s – maybe because of Balthus. And then, from her family house, she dug out lots of original women's clothes from the 1930s, and she put them to good use…

VH: In particular, the shoes. All along, there is a contrast between white and black shoes.

JR: It is possible, but on this topic you know better than me.

VH: So the scenes you are unhappy with are mostly in the part that takes place in the mansion…

JR: Yes, because then Monsieur Berta took even longer than we had reckoned for the lighting of each scene. So we were under tremendous pressure, especially on the last days when it became essential to shoot all what we could. But again, I haven't seen the movie since. And if I saw it now, I really don't know what I would think of it.

VH: This movie is extremely powerful. It is shorter than your other movies, and also has a lot of 'punch'.

JR: Since it was necessary to condense quite a lot, by force of circumstance, I believe that it is indeed the most elliptical of all my movies. Otherwise I might have made a three or three-and-a-half hour movie, like I usually do. But there, we were obliged to simplify, to keep to the essentials. It might have given a more vigorous and energetic feel to it…

VH: Regarding the end, it looks like all the adaptations of *Wuthering Heights* have suffered from great production constraints: Wyler wished the end had not been imposed on him, Tilley was thoroughly disgusted by the way the editing was done.

JR: In contrast, Buñuel's end is superb.

VH: In the tomb…

JR: Completely invented. It is extraordinary… It is beautiful, but different, since the two main actors are definitely 40. Besides, amongst all the versions I know, Buñuel's is the 'shortest': he uses the most compressed timeline. Not only does he stick to the first part, but he also starts when the adult Heathcliff comes back, and finishes with Catherine's death, which represents a quarter of the novel.

VH: After all, Brontë's novel had to be divided into two parts, two volumes, to be saleable.

JR: Still, it does not count. The second part of the novel is equally beautiful, superb…

VH: Truffaut kept the same actor, Jean-Pierre Léaud, in *The 400 Blows* and *Love on the Run*, the last movie of the Antoine Doinel cycle, with an interval of 20 years between the two films. If you had had the same opportunity with Lucas and Fabienne, would you have been tempted to make the second part of *Hurlevent*, a generation later?

JR: Not with Fabienne since Catherine is dead. So another actress would be needed. But with Lucas, it could have been interesting. But I won't do it. Anyway, now, it is too…well no, it has been eighteen years since 1984.

VH: And when was the movie released in the theatres?

JR: If I remember correctly, it was released in Autumn 1985, in September or October 1985. So, at the time, I thought to myself that it was a strange idea to choose and release it in September, when there were already loads of other movies being released concurrently.

VH: Who decides the release date?

JR: Not me, the distributor does. Besides, the distributor was linked to Berri, so they decided… But it is not their fault if the movie was a flop. This movie belongs to a category of movies that people like or do not like. There are movies that, more or less, appeal to many people for many different reasons, like *Va savoir* (2001). On the contrary, there are movies that some people like a lot while, to others, they mean absolutely nothing. For example, *Up, Down, Fragile* (1995) belongs to this category. It is a movie that people either like a lot, or, they cannot even conceive why we thought of making it in the first place. And it is going to be the same story with *Marie and Julien*. This I know in advance – whether it is good or not, some people will love it and others will hate it.

VH: Do you think that this has to do with their 'reading', or interpretation, of the movie?

JR: It has to do with the topic, though I could not say exactly why. That's the way it is, there are some movies that really divide people. Those who are interested are even more interested; however, those who are not cannot even start to imagine what triggered the idea. It was fairly noticeable with *Up, Down, Fragile* and will certainly be with *Marie and Julien. Hurlevent* worked much the same way but, on top of that, people have pre-conceived ideas about *Wuthering Heights* which has coalesced in their minds with the Wyler movie. But, for me, Wyler's movie is vaguely faithful to the novel, in the letter, but makes no sense whatsoever with all those ball scenes sprinkled everywhere. In fact, they transformed it into an 'Emily Brontë and Jane Austen' production. Actually, *Wuthering Heights* is Wyler's movie, after a novel by Jane Austen!

VH: But do you really think that the Wyler movie is being so vividly remembered by today's true cinéphiles?

JR: No, but the only interesting thing is that Gregg Toland did the photography. Therefore, thanks to him, there were moments that were visually powerful... Laurence Olivier, of course, is a fabulous actor: had he done it ten years earlier, he would have been fantastic.

VH: Adapting Emily Brontë's novel is not a walk in the park and, nowadays, in 2000, there are more adaptations of Jane Austen's novels than of *Wuthering Heights*. So why this desire to make a film based on Brontë's novel since most of these ventures fail one way or another – commercially, mainly? During the shoot, there is always something that does not happen or does not happen according to plan. Or else, it is very difficult to find the right actors.

JR: There have been some successful film adaptations of Jane Austen's novels, though?

VH: Yes, Jane Austen, but...

JR: I don't know her novels so well, but there is this adaptation with Emma Thompson in it...

VH: *Sense and Sensibility* (Ang Lee, 1995)?

JR: It wasn't too bad, was it? It is not a great movie because it is treated in a very impersonal manner…

VH: With Jane Austen, it seems to be easier.

JR: Well yes, because first of all, the dialogues are already completely written: they just need shooting. Jane Austen was a true genius when it came to writing dialogues.

VH: Whenever a good version of *Wuthering Heights* emerges, the film-maker takes a particular stance, emphasises a certain angle, a personal vision – especially for the character of Nellie, the servant. Therefore, the fact that you did not re-read the novel makes sense to me.

JR: I don't know. I felt it was necessary not to re-read it straight away. I re-read it later on, though. In fact, I started to re-read bit by bit, when we had decided on everything, when we already had our scenario. Then, I re-read it completely. And afterwards, I was able to say: 'What about using the sentence she uttered at that moment – or this sentence, or that one'. Besides, on re-reading it, I could not help thinking that it was a superb piece, the work of a genius.

VH: And did you try to re-read it in English too?

JR: No, I did not. I was just curious to see how it was translated. I cannot read well enough in English. I can read newspaper articles in English, but cannot read English fluently – in particular, when it is a text as well written as Emily Brontë's. I did try with another Anglo-Saxon author – not an English one – who is very badly translated into French, James. It is indeed very difficult to translate him. Half to three-quarters of the translations of James are awful in French. It so happens that I have compared the translation and James' original text a couple of times. In the French translation of *The Wings of the Dove*, there are some passages that do not mean anything… On following the English text, I realised that the translator had translated word for word because she did not fully understand. Therefore it was a translation that did not mean anything in French. Conversely, there were things she had over-simplified, like the first sentence: 'She waited, Kate Croy,' which is so beautiful but ended up being completely flattened out. She simply wrote: 'Kate Croy attendait son père'. In short, all the syntactic effect had been totally destroyed…

VH: All the visual quality of the interpolated clause…

JR: Yes, the particular syntax and, also, the polysemy…

VH: Similarly, in your movie, one thing can have several meanings and this leaves a lot of room for interpretation. But who got the idea of the central scene – the oneiric scene?

JR: I cannot remember… It was meant to represent a division… It felt obvious, necessary for this oneiric scene to be in the middle and materialise the passage of the three years during which Heathcliff is absent. I simply felt like having this scene which, indeed, is a creative add-on.

VH: And what about the powerful idea of the blood that appears progressively, like some stigmata…

JR: I forget what we did. I just know that we came up with the idea in an attempt to represent the passage between the two parts, since the first and second parts are about the same length. We knew we needed a visual equivalent, so we felt we could have an oneiric scene – like we had at the beginning and…

VH: At the end…

JR: This is a principle that I have applied again to *Marie and Julien*. In fact, I realise now that *Marie and Julien* has been very much influenced by our adaptation of *Wuthering Heights* since the project was born two years later.

VH: Do you wish to talk about *The Story of Marie and Julien*?

JR: No, not particularly. I only wanted to mention that it had been structured around dreams like our adaptation of *Wuthering Heights*. There is a dream at the beginning, one in the middle and one at the end. It does not end with the dream, even though the audience will no longer know when it is a dream and when it is reality.

VH: In *Hurlevent*, I didn't realise that the first scene was a dream. To me, the time just seemed to be suspended. It is thanks to Pascal's interview, which is present on the DVD collector pack released by ARTE, that I learnt about it. However, it is the dream scene located in the middle of the movie that makes everybody want to see the film.

JR: As far as I can remember, I was quite happy about it.

VH: How did you come up with the creative idea? Did it simply occur to you?

JR: Well, I don't know. I think that we talked about it (the three of us), that we sat down and had a discussion… Personally, I like working that way: we write very little. It is Pascal who writes, afterwards. But apart from that, when we see each other, we talk, we talk, we talk and then, sometimes, the good ideas come. Suddenly, you say something that triggers a reply. And that's why I quite like working in a trio. Because when three people have a discussion, they hand over the baton frequently whereas two people often go round and round with their opposite views. In a trio, the third person takes sides for one or the other. Additionally, clearer arguments have to be given. That's how you move forward. It has worked the same way for *Marie and Julien* with Christine and Pascal. We also had loads of problems – for instance, finding the end that did not exist in the text I had dug out – and tackled them by discussion.

VH: Regarding the release of *Hurlevent*, there were apparently a couple of hitches. I have the feeling that its screen life was cut short in the theatres.

JR: No, it could not be different because the audience was not there. So, when there are only a few spectators, the movies cannot be shown in empty theatres for long. That's why I am happy about the DVD. Now, there are a lot of people who will see it who were not even aware that the film existed. They may well buy the DVD for *La belle noiseuse* or *The Gang of Four* which are more popular… But then, they will feel like watching the other movies.

Yoshida–Bataille (1988, Japan, 125 min)

It took Luis Buñuel twenty years to realise his Surrealism-tinted *Abismos de pasión* in Mexico, and Kiju Yoshida thirty to materialise his version of *Les Hauts de Hurlevent*, a costly 125-minute Jidaigeki, transposed at the end of the Kamakura era, in the fourteenth-century Japan. To date, it remains Yoshida's only Jidaigeki, or rather 'anti-Jidaigeki' or *anti-period drama*, since *Wuthering Heights* 'does not correspond to a [Japanese] national itinerary'.[405] The Heian and Kamakura periods, stretching over five centuries of medieval Japan, were marked by the rise and fall of the Shoguns, and had previously inspired Kenji Mizoguchi's *Sansho the Bailiff* (*Sansho Dayu*, 1954), the heart-rending story of two noble children and their mother, who are sold into slavery. Mizoguchi's return to the Imaginary of the remote Heian Japan had given him the liberty to deal with raw emotions, and would give Yoshida the same kind of freedom in 1988. As Yoshida tells Mathieu Capel, in the introduction that features on the DVD of *Onimaru* (2009), the period atmosphere of *Sansho the Bailiff*, the sheer intensity it lent to its subject was what he always wanted for *Heathcliff the Devil*. He would put back Emily Brontë's archetypal heroes, Heathcliff/Onimaru and Cathy/Kinu the Mother, where they really belonged, at the very bottom of the three-dimensional Gothic architecture of *Mise en Abîme*, in the semiotic and ontophanic domain of myths, and only Hareton/Yoshimaru and Catherine/Kinu the Daughter would eventually emerge from its depths.

Since his début at the *Shochiku Studios*, and all through his apprenticeship there, from 1955 to 1960, as assistant director to, principally, Keisuke Kinoshita, Kiju Yoshida had always been involved in the literary phase of scriptwriting. Yoshida, who had co-founded with his peer, Nagisa Oshima, the magazine *La Critique de Cinéma* (1960), which drew its inspiration from the French and Japanese New Waves, believed that he could only live his passion for cinema as an 'auteur' who chooses his literary source material and adapts it himself. In 1985, on resuming his international career after a five-year immersion into the Mexican cinema and television industry, he made his feature film, *A Promise* (*Ningen no Yakusoku*, 1986), while hoping for the green light of *Seibu ryûtsu gurupu* to start the film project based on his completed script of *Onimaru*, his own cultural translation of *Les Hauts de Hurlevent*. Yoshida conceived *Onimaru* as a tribute to Buñuel, whose presence was still palpable in the artistic circles of the Mexico City of the early 1980s, and whose *Wuthering Heights/Abismos de pasión* had been revived by some tribute screenings in New York, following his death in the Summer of 1983. Owing to the commercial failure, in Japan, of Akira Kurosawa's *King Lear*-inspired *Ran* (1985), the production of *Onimaru* (also entitled *Arashi ga oka*) was postponed for two years. Eventually, with the additional financial support of *Mediactuel* and Francis von Büren, came the opportunity for Yoshida to realise his film on the barren slopes of Mount Fuji, at a spot known as Gotenba. The director, during his two-year wait, had gained the support and expertise of *Ran*'s production designer, Yoshiro Muraki, and *Ran*'s music composer, Toru Takemitsu. Their names had probably reassured the Western investors.

Muraki, the production designer, conforming to the tradition of the Jidaigeki, built from scratch the fortified manors of Higashi no Sho (Wuthering Heights/The Eastern House), and Nishi no Sho (The Grange/The Western House), which immediately filled the backdrop of the First Mythical Component (MC1). Furthermore, the Yin and the Yô of the Japanese ancestral beliefs came to be superimposed on to the paradigm of the Noble Houses of Yamabe that are situated on two opposite slopes of the same awesome volcano, also called Mountain of Fire. Its destructive and re-generative power is harnessed by the Lords of Yamabe, through the rituals of the Serpent, which consist in the passage, at nighttime, under a multitude of Torii gates (symbolising the slough of the Serpent) and culminate, for the elected Lord, in a primitive Kabuki dance, ending up in the trance of the God of Fire. The Lords of Yamabe belonging to Higashi no Sho, the Eastern House, comply with the rituals in a fanatic and war-like fashion, while the other branch of the Yamabe family living at Nishi no Sho, the Western House, proceeds by meditation, offerings and prayers, and includes the Yamabe women in the ceremonies. Nonetheless, as explained by Yoshida to Josiane Pinon-Kawatake, in the interview published in *Positif* in 1987, the Yin and the Yô thus incarnated on screen emphasise the 'utter cultural isolation' of the two noble houses, rather than any 'permanent moral antagonism' between them.[406] As well as being a challenging cultural translation for the Western viewers, *Onimaru (Arashi ga oka)* represents a linguistic hurdle for the non-Japanese speakers, who need to rely on translators, for the subtitles, and bilingual film critics, for the interviews, in order to make sense of the complex film images, which Yoshida intertwined with dialogues in most cases and shot, always, with Takemitsu's soundtrack in mind. Moreover, for the majority of viewers, those who had not seen the film in a cinema when it was first released in 1988, the unavailability of subtitled video tapes in the usual European and North American commercial circuits made necessary, up until March 2009, the involvement of film institutes and festivals – as well as the collaboration of local producers, namely Philippe Jacquier, in France – to view a print, or avail of a video copy with the requisite subtitling. This major difficulty of finding a copy of *Arashi ga oka* in a Westernised subtitled version has hampered academic appraisal itself. So far, only Kamilla Elliott, in 'Literary Cinema in the Form/Content Debate' (2003), the chapter using *Wuthering Heights* as a literary specimen for 'rethinking the Novel/Film debate', takes the risk of a critical integration of the film into her argument.[407] This judicious integration is motivated by the showcasing of her 'Genetic Concept of Adaptation', and leaves out the creative history and translative merits of the film. Her genetic concept reveals an a-version from *Arashi ga oka*'s Third Bataillan Theme (BT3) which leads to some slight interpretative mistakes, adverse to her argument, in most cases. In particular, Elliott's misappraisal of the dramatically effective mirror scene where Kinu/Cathy, pregnant with Onimaru's child, declares, 'Kinu is Onimaru. Onimaru is Kinu. Onimaru is myself', is typical – in the words of George Steiner – of the '*appropriative* comprehension' of a source text by scholars who 'disrupt' the text, without allowing it (or its auteur/author) to re-arrange their own.[408]

Prompted by the Yoshida rétrospective at the Centre Georges Pompidou, in Paris, in the Spring of 2008, *Onimaru/Arashi ga oka* finally became accessible, a year later, in a digitalised and Westernised version (with French subtitles only, though). This DVD, put together by *Carlotta Films* thanks to the input of translator and film critic, Mathieu Capel, is essential as it contains the first interview of Yoshida on *Onimaru/Arashi ga oka* since Cannes 1988. The title of the film itself is a lexical mystery that is worth deciphering. It translates into 'Oni', 'Devil' and 'Maru', 'Circle', which seems to suggest that Onimaru, an evil character comparable to Mizoguchi's Sansho, is and will be at the centre of the narrative for this, and the next versions-to-come of 'anti-Olivier' interpretations. Still, 'Arashi ga oka', the original Japanese title, translates into 'Arashi', 'Storm' and 'ga oka', 'Mount' – the Mountain of Storms, or Wuthering Heights. Thus, Yoshida's film, like Brontë's novel or Wyler's movie, recounts much more than the story of the evil Heathcliff. The eponymous hero, whose evil springs from an impossible quest for recovering his identity, is infinitely more closely related to Mizoguchi's oppressed orphan Zushio, than to Sansho the Bailiff. In the first sequences of the epic, *Arashi ga oka*, Takamaru/Old Earnshaw introduces the dishevelled orphan, Onimaru, to his children, Kinu and Hidemaru/Hindley, as their new brother, then chooses him over his own son for his physical courage and divinatory abilities, to take the lead in the Ceremonies of the Serpent. Despite Takamaru's arbitrament in his favour, Onimaru's position always seems precarious within the caste, and his identity as Lord of Yamabe at Higashi no Sho, after Takamaru's death, usurped in regard to Hidemaru's birth rights. After a long absence, returning to the Mountain of Fire with the title of Lord of the Two Manors, conferred on him by some provincial Shogun for his services as a ruthless samurai, Onimaru remains no more than a ruthless outsider. His status is dubious because of Mitsuhiko/Edgar's ancestral rights as Lord of Yamabe at Nishi no Sho. Further, Mitsuhiko has agreed to marry Lady Kinu of Yamabe, which honours him with the marital legitimacy of the former Lords of Yamabe. Hence Onimaru's wrath, and his failure to recognise that Tae/Isabella, as Lady Tae of Yamabe, is not merely following him to Higashi no Sho out of self-interest, but is also hoping that he will accept from her, the same kind of legitimacy that Mitsuhiko has accepted from Kinu.

In *Arashi ga oka*, Onimaru's quest for identity is, of course, indissociable from Kinu's. From childhood, Kinu has endeavoured, by the means of an elaborate role-play involving a precious circular mirror, to resolve the mystery of her own identity through Onimaru (since she has no mother), and to find her righteous place (as well as Onimaru's) within the Yamabe religious caste. At Higashi no Sho, the Yamabe caste condemns to exile its female descendants and, in the absence of mothers, can exclude a legitimate son, while not ever wishing to include a natural one. Thus, Yoshida, if one considers the new insight that his cultural translation gives into the novel's Gothic Figure of Lineage, can be said to *compensate* for the '*appropriative* comprehension' of Emily Brontë's text into his Jidaigeki.[409] Yoshida goes even further in the intellectual reciprocation with his translation of the Bataillan Themes (BTs) and, especially, of the Second (BT2; Hyper-Morality and Morality) and Third (BT3; Sexuality and issues of Physical Death and Re-Generation) ones. This is hardly surprising as he declares himself, in the introduction to the 2009 digitalised version of *Onimaru*, that Georges Bataille's essay on Emily Brontë, in *La littérature et le mal* (1957), had first driven him to read *Les Hauts de Hurlevent*, which, he believed until then, was 'destined for the education of teenage girls', then spurred him on to translate it into a Jidaigeki. *Les Hauts de Hurlevent*, he agreed with Bataille, was an 'a-moral tale' touching on the rapports between the Profane and the Sacred, and instancing an 'a-moral' or rather, in the words of Bataille, a 'Hyper-Moral' case of Transgression:

> The tragic author [Emily Brontë] agreed with the law, the *transgression* of which he described, but he based all *emotional impact* on communicating the *sympathy* which he felt for the *transgressor* [Heathcliff]. [...] p. 9

Such *temporary transgression* is all the more *free* since the *interdict* is considered *intangible*. So Emily Brontë and *Catherine Earnshaw*, who both appear to us in the light of *transgression and atonement*, depend less on morality than on hypermorality. *Hypermorality is the basis of that challenge to morality which is fundamental to Wuthering Heights.*[410] [Emphasis added] p. 10

Antoine de Baecque from *Les Cahiers du Cinéma*, in his article entitled 'Gods and Men', associates the divine, the imposture and the interdicts (part of BT2) with the stern faces, tattooed bodies and heavy hieratic postures of the Lords of Yamabe, at Higashi no Sho.[411] In provoking his bloody Chambara (Japanese sword fight) with the loving and invincible Yoshimaru, Kinu the daughter, who embodies the humane and the truth, transforms Onimaru, the Lord of the Two Manors, into a maimed and defeated half-god, and forces him to wander on this narrow strip of no-man's land separating the Profane from the Sacred. In pushing her lost father there, and in resurrecting Onimaru, the cursed orphan, she puts into question the interdicts, exclusions and executions imposed by the Yamabe religious caste, and by the military rule of the Shoguns, which Onimaru had come to embrace out of necessity and revenge. On his return to the Mountain of Fire, having realised that part of his identity lay in the forsaken Kinu, he had wreaked havoc on the Yamabe Houses; exiled on this narrow strip of land where he first travels with Old Takamaru as a dishevelled orphan, he finally realises that he can heal his wounds, reach the lost Kinu, and allow her to find her way out of Purgatory and into the Valley of Buddha. To de Baecque, the lightness of being, laughter and graceful bodies that 'carry the fragile humanity' belong to the Ladies of Yamabe who, from Kinu the Mother to Kinu the Daughter, and from Shino/Frances to Tae/Isabella, progressively manage to detach themselves from tragedy and revenge. Effectively, the weapon of the younger generation of women has two sides; Language as a means of deriding the men-monks, who believe in their elemental birth from the entrails of the Mountain of Fire, and recovered Identity as a means of asserting one's existence in a lineage of noble women, independently of the side of the Mountain on which they are born:

Onimanu is made to come back to Earth with Kinu, stumbles on Tae and is executed by Kinu's daughter.[412] [My own translation]

In *Arashi ga oka*, the Third Bataillan Theme (BT3) is not mitigated by the parallel experience of a sophisticated intra-diegetic observer since what needed to be conveyed directly to the viewers, the extra-diegetic observers, was the archaic, demiurgic stage when life appeared on Earth. Yoshida, in that same preface to the 2009 DVD of *Onimaru*, quotes immediately from Bataille's essay on Emily Brontë, 'Eroticism is the approval of life up until death', and explains that he had re-imagined, for Kinu and Onimaru, the undifferentiated stage of the embryonic cells when they still have the potential of being both male and female; the cells of Kinu's and Onimaru's whole beings seem to be called back to that early reproductive stage. Therefore, the eroticism in the film has nothing to do with pornography; it runs deeper in the subconscious, at the sub-level where the artist draws her/his creative forces, and where images and words are still in the making. De Baecque's intuition was to interpret Onimaru as the Trespasser and Learner of Yoshida's *Arashi ga oka* since, as an invincible samurai, he is being defeated by 'the humanity present in another body'. Onimaru is first challenged by his encounter with Kinu's flesh in its youth, which is highly eroticised in the love scene of the Forbidden Chamber, then overwhelmed by his awful discovery of Kinu's flesh in a state of decay, dehumanised by Death, at Jobutsu ga Tani, the graveyard hanging on the ashy slopes of the volcano. Last but not least, this Third BT (BT3) is incarnated by Kinu and her extreme experience of giving birth to another Kinu. Her deathly duplication exemplifies vividly, on a cinema screen, what Georges Bataille was summing up in the first paragraphs of his unavoidable essay:

> Sexuality implies death, not only in the sense in which the new prolongs and replaces that which has disappeared, but also in that the life of the being who reproduces itself is at stake. To reproduce oneself is to disappear, and even the most basic asexualised being is rarefied by reproduction. [...] Individual death is but one aspect of the proliferative excess of being.[413]

In the first sequence of *Arashi ga oka*, an unidentified young man, blind and playing the shamisen, returns to the Mountain of Fire after having trained as a monk in the capital. At Nishi no Sho, he asks a very old spinner (maybe, Osato/Ellen) about *Lord Onimaru* of Yamabe and *Lady Kinu* of Yamabe. He appears to be either a brief instantiation of Lockwood, or a dramatic upgrading of the little shepherd boy featuring modestly at the very end of Emily Brontë's novel, since all he learns about Kinu, after having trespassed the limits of the graveyard, is that her tomb is guarded by a proud and ruthless horseman, who does not hesitate to slay the desecrators with his awesome sword. Yoshida's Jidaigeki version of *Wuthering Heights* is not meant to accommodate the narratorial players living at the extreme top of the novel's Gothic architecture of *Mise en Abîme*. The temporal (and spatial) displacement to the Kamakura era corresponds to a translative panacea, which is designed to *do away* with them. The only flaw (or narratorial specificity) thus, as regards the dynamic structures of retelling characterising this powerful translative strategy, is that, without the intermediary level of the intra-diegetic outsiders, Kinu and Onimaru, like the Catarina and Alejandro of the Buñuel version, remain stuck, forever, at Jobutsu ga Tani (the Valley of Buddha), in the inhospitable domain of myths, leaving Kinu and Yoshimaru to struggle to remain in the memories of the spectators, on their way back to the more civilised side of their Mountain of Storms, Nishi no Sho.

Philippe Jacquier (*Onimaru*, 1988)

Philippe Jacquier and Kiju Yoshida

Philippe Jacquier is at the head of Sépia production, a film production company based in the 21ˢᵗ District of Paris (France). Sépia co-produced the latest movie of the highly esteemed Japanese film-maker Kiju Yoshida, Women in the Mirror (Kagami no Onnatachi, 2001).

Women in the Mirror was selected at the Cannes Film Festival in 2002, twelve years after Yoshida's previous feature, Onimaru (Arashi go aka, 1988), a superb transposition of Wuthering Heights at the end of the Kamakura era.

Sépia Production, Montreuil (Paris), 23ʳᵈ Feb, 2004

VH: The documentary *Cinema Dreams, Tokyo Dreams* which Yoshida made in 1995 revolved around the character of your great-grandfather, Gabriel Veyre.

Was it through this documentary that you discovered Kiju Yoshida? Or had you heard of him before?

In which circumstances did you then meet him?

PJ: The documentary came much later. The first time I heard of Yoshida was at the Cannes film festival back in 1988. I saw *Onimaru* as a simple spectator – I was not a film producer yet – and was quite amazed by his work. After that, I felt like learning a bit more about Yoshida, so I was on the lookout for his films at the *Cinémathèque Française*. Very few of his films were actually shown there, but I managed to see *Eros + Massacre*[414] soon enough.

On a more personal level, it was around that time that I discovered that my great-grandfather was once a cameraman in the company of Louis and Auguste Lumière. I had unearthed a correspondence between my great-grandfather and his mother which spanned four years of travel around the world. So I decided to follow in Gabriel Veyre's footsteps. This was roughly around the same time in 1988-1989.

So I went to Mexico, to Cuba, to Canada and to Japan. When I travelled to Japan, I took with me the videotapes of the films that he had shot there. He would often be the first film-maker in those countries that he visited. In Japan, to be perfectly accurate, he was the second one – the second operator to shoot some films for the Lumière Brothers.

When I arrived in Japan, I sought to meet people who could be interested in his films. And this is how I met Shigehiko Hasumi who, to me anyway, is one of the greatest Japanese intellectuals. Hasumi is both an intellectual – he was Dean of the University of Tokyo ('*Todai*') for two years – and a cinéphile. He wrote on Jean Renoir, John Ford and Yasujiro Ozu. His book on Ozu was published in France by the *Cahiers du Cinéma*. He speaks French like you and me, he knows *Madame Bovary* by heart. No sooner had I met him that he took me to the libraries where we could dig up the articles of the *Japan Times* dating from the time of Gabriel Veyre in Japan. Gabriel's letters helped us to map out his itinerary – he would write: 'I did a screening in this or in that place' – and, thanks to them, we were able to make some significant findings in the newspapers. Subsequently, Hasumi told me: 'There is a film-maker who could be interested in your project; Yoshida'. Since I was discovering his films at around the same period, this struck me as a remarkable coincidence. Before *Onimaru*, I would have been more inclined to meet either Shohei Imamura or Nagisa Oshima but, at this stage, Yoshida's cinema interested me deeply. And Hasumi concluded: 'On top of that, he speaks French!'.

VH: Did you find all these letters while rummaging in an old attic?

PJ: In an old family wardrobe! And I did not only find these letters, which were annotated, but also some glass plates. Gabriel's letters read like this: 'Dear Mum, I am taking my Lumière cinematograph to Mexico...'. I had unearthed all the correspondence he had written over his four years of travelling around the world for the Lumière Brothers from 1896 to 1900. Then, from 1900 onward, would you believe it or not, he became the official photographer and film-maker of the Sultan of Morocco! But my book on Gabriel Veyre stops in 1900.[415]

VH: Yoshida studied French literature, didn't he?[416]

PJ: And he loves France. At that time – I think it was in 1990 – he had just staged *Madame Butterfly* at the Opéra of Lyon and I really felt like meeting him. So the first time I met him was in Saint-Germain-des-Prés, at the *Café de Flore* or at *Les Deux Magots*, not in Tokyo. On that first meeting, I was quite impressed because he was right in the middle of restaging *Madame Butterfly* for Grenoble. I told him about the story of Gabriel Veyre and he watched the films. Between us was born a kind of teacher-student relationship, which also became a friendship. He believed he could make a feature out of the story of Gabriel Veyre. He was quite taken by his story.

In the meantime, I had become a producer – earlier, I had worked in the film production business but did not have my own company. As a friend and as a producer, I went to Grenoble, saw Yoshida's *mise en scène* of *Madame Butterfly* – which was fantastic – and fell in love with it. The first documentary that Sépia production produced was entitled *Kiju Yoshida meets Madame Butterfly*. You see, his work was already linked to the theme of the atomic bomb: the setting of *Madame Butterfly* in the streets of Nagasaki had a deep meaning.[417]

Then we started to work together on Gabriel Veyre – on a feature film that simply took time not to materialise! Yoshida spent three to four years writing a scenario and often came to stay in Lyon. I also travelled to Japan once or twice. When came the centenary of the Lumière cinematograph, we still had not managed to shoot the feature on Gabriel Veyre. So we had this idea of making a documentary for a Japanese television channel. That's how Yoshida shot *Cinema Dreams, Tokyo Dreams*, a documentary about Gabriel Veyre, a character he knew by heart.

VH: For which television channel was it made?
Was this documentary followed by another collaboration?

PJ: The documentary was made for Metropolitan TV, a television channel which had just been launched back then in 1995. But it should be possible to see it soon. I am hoping for *Madame Butterfly, Cinema Dreams, Tokyo Dreams* and *Women in the Mirror* to be all three digitalised and released by Sépia in a DVD collector's box by the end of the year![418]

After the documentary *Cinema Dreams, Tokyo Dreams*, it is true that we wanted to continue working together. Since Yoshida had not made a feature film since *Onimaru*, I really wanted to create the opportunity for him to shoot one again. Therefore, I bought the copyright of an English novel, a novel by Kazuo Ishiguro, *A Pale View of Hills*.[419]

VH: What happened with Ishiguro's novel? Has Yoshida ever adapted it?

PJ: The meeting between Ishiguro and Yoshida in London went beautifully well. When we left, Ishiguro had given us the copyright of his book for next to nothing. And despite the fact that the project was quite complex to put together, I was quick to set up the production. But the day before the shoot should have begun – this is reality, not fiction – the film project stopped. The sets had been ready on time. The casting was fantastic with Mariko Okada[420] in it. Everything stopped because of an argument, a big production argument – not very pleasant to recount, I am afraid – between the English, the Scots and the Japanese. It stopped and I was forty. I lost all the money I had put in Sépia and thought that Yoshida might not be willing to put his trust in another feature film again. Although he was keen on the Ishiguro adaptation, I had been the driving force behind it and had also served as a point of contact – Ishiguro being both Japanese and English. I did wonder if Yoshida would feel up to having another go at *A Pale View of Hills*. Therefore, I tried to put together the project again with some Canadians. It did not materialise either. And, as I was still putting all my energy in that project, I lost two or three years' worth of production. So, in many ways, it was quite painful. But, what was positive is that we stayed very much in touch. And he called me back one year with a new project in mind: 'I have a new idea, this is not an adaptation of Ishiguro, it is more like me'.

So, thanks to this rather bad experience, we aimed at something else, something which was a hundred per cent Japanese, and I wanted to be the co-producer of this new feature. I raised the money in France and managed to co-produce *Women in the Mirror*, a superb movie, which was selected at Cannes in 2002. I believe it is one of Yoshida's most beautiful pictures. Besides, he had not shot with Mariko Okada for thirty years: it was their twelfth film together.[421]

VH: Who distributed *Women in the Mirror*?

PJ: Les Films du Paradoxe. But they have nothing to do with the production. They appear on the film poster but, in actual fact, 'Sépia production presents'. It was me who brought them the film once it was completed.

At the time being, I am preparing the collector's DVD with *Women in the Mirror* and the two documentaries. Since none of his movies are available in DVD, we are also thinking of releasing *Eros + Massacre* on its own. There are two versions, a three-hour-and-thirty-minute version, which is a bit tough, and a two-hour-and-forty-minute version, which is more digestible.

VH: Between *Onimaru* and *Women in the Mirror*, he made the documentary on Gabriel Veyre. What else?

PJ: He also made a documentary on Yasujiro Ozu and wrote a book, *Ozu's Anti-Cinema*, which was published in the United States[422] in 2003 and will be in France within two months, in the *Institut Lumière/Actes Sud* series. He presented this book at a symposium dedicated to Ozu which took place in the United States in November 2003.

VH: The relation between Yoshida and Ozu is interesting – and quite ambivalent really.

PJ: In fact, Yoshida came across as being hostile to him when he was a young film-maker. They used to work in the same studios.[423] Yoshida was then the assistant of Keisuke Kinoshita. In a nutshell – and it is the beginning of his book – Yoshida had been critical of one of Ozu's films.[424] Ozu never really bore a grudge against him. Still, there was some kind of tension between them. Interestingly enough, Yoshida's wife, Mariko Okada, played in Ozu's last movies,[425] and her father[426] had played in almost all his silent films.

VH: Regarding *Onimaru*, what had made such an impression on you?

PJ: I saw it in competition at Cannes when I did not know the first thing about Yoshida. I saw it as a Japanese film. I have always been attracted to foreign films – and it is quite understandable when you consider that Gabriel Veyre was my great-grandfather! What I remember about *Onimaru* was its powerful story and its powerful rendition – the extreme rigour in the photography, in the composition of the image, in the direction of the actors. I thought that *Onimaru* was not only aesthetically powerful but that it also conveyed some important meanings that I could not necessarily grasp fully – as exemplified by the motif of the mirror and the double which is recurrent in his films. I also found that there was a very powerful eroticism in *Onimaru*. So I said to myself: 'Well, here is a great film-maker and you don't even know him!'. And it was his eighteenth film!

Once I had seen the movies he made in the 1960s and in the 1970s – after *Onimaru*, I saw everything that could possibly be seen – I realised that he was a virtuoso of the *mise en scène*. When I talk of *mise en scène* in cinema, Yoshida's name comes first to my mind. I believe that people do not know any longer what *mise en scène* really is about. They give *mise-en-scène* awards indiscriminately. In Yoshida's films, the *mise en scène* reflects a deliberate artistic choice. For instance, he could be waiting for the right light for hours. And he knows better than anybody how to organise abstractly a series of shots. When he shoots a film, he is well capable of taking as many as thirty shots in one day. These shots would have been conceived and ordered precisely in his head. In Yoshida's cinema, nothing is done by accident. This, I find extraordinary.

Besides, he is a serious person and there is no room for giggles in his cinema. His films always trigger some kind of reflection and do not possess the lightness that most people love. I do like films that make you think. On starting a new project, Yoshida always makes a very simple and radical artistic choice, then relies on the power of the metaphors – like the metaphor of the mirror. In my opinion, Yoshida is also the film-maker who has best filmed women in Japan, and it is this aspect of his cinema that interests me

291

most. He has dealt with some powerful subjects, making women the central characters, and they are absolutely magnificent in his films. His wife, Mariko Okada, has demonstrated it best. In comparison, Akira Kurosawa is as great a film-maker but his cinema is about men. And the remakes of his movies are masculine, produced in Hollywood as Westerns. But Yoshida has a more feminine approach to film-making, which reveals itself in his *mise en scène*, in his subjects and in the choice of his actresses. Of course, I am thinking of *Akitsu Springs*[427] but there are also, in close succession, *A Story Written on Water*,[428] *The Lake*[429] and *The Affair*[430] – which are first and foremost women tragedies. When he stages *Madame Butterfly*, it is ***Madame*** Butterfly. When you watch his nineteen feature films, his theme, his only theme, is woman. And *Wuthering Heights* is no different.

VH: In both Emily Brontë's novel and Yoshida's adaptation, Heathcliff/Onimaru appears to be the central character but, in fact, Heathcliff's story belongs to the story of two generations of women, the two Kinus (Cathy/Catherine), mother and daughter.

I was impressed by the way the first Kinu acted and was filmed. There is this seduction scene with the fans between Kinu/Cathy and Mitsuhiko/Edgar. Their body language totally blends into the *mise en scène* and it is impossible not to remember that Yoshida is also a stage director.

When Kinu first arrives at Nishi no sho (The Western House) in her rich bridal attire, Mitsuhiko must be struck by her unearthly appearance. While Tae/Isabella makes her kneel comfortably in front of the fire, Mitsuhiko sits sideways and takes up his fan. As Kinu unveils the reason of her distress, he is soon taken by her exquisitely slow gestures and her soft voice coming forth from behind the interstices of her fan. For the remainder of their interview, Mitsuhiko then lowers his fan and turns his whole body towards her. Quite demurely, Kinu slowly lowers hers, revealing her face to his eyes.

Follows the deeply erotic choreography staged by Yoshida, in the Forbidden Chamber of Higashi no sho (The Eastern House), where Kinu has invited Onimaru. The taboo of a passion going against the rules of society and defying the tangible presence of death itself seems to have been lifted, for a brief moment, in that Forbidden Chamber. Yoshida, I believe, said he had been profoundly inspired by Georges Bataille's essay[431] on Emily Brontë in *Literature and Evil* (1957).

PJ: Yoshida's cinema could be qualified as highbrow cinema, even if *Onimaru* (or *Women in the Mirror*) can touch people who are not necessarily cinéphiles. His cinema requires from the audience some kind of intellectual work. All the keys are not given on the first viewing.

Incidentally, Peter Kosminsky – who directed the adaptati
Heights with Juliette Binoche[432] – had also wanted to adapt
Isn't it quite funny that two film-makers who have apparently nou...
common should wish to adapt those two same novels, *Wuthering Heights*
and *A Pale View of Hills*. But Yoshida knows what it would imply to make
A Pale View of Hills, Kosminsky does not. It is not possible. A non-Japanese
who deals with Hiroshima, it is not possible. Well, Alain Resnais[433] did it…

*   *   *

The Japanese version by Kiju Yoshida is an intensely sophisticated cinematic rendition of
Emily Brontë's novel, which adds considerable aesthetic and academic value to its source
material. It follows an extreme yet well-wrought adaptative strategy, which has helped
us in apprehending better the functioning of all adaptative (or, to use Steiner's epithet,
'translative') schemata applied to *Wuthering Heights*. The height of artistic achievement
for independently minded film-makers and producers, Yoshida's *Onimaru* serves
as a reminder of the continuing fertility of the text and of its unconscious, abroad, in
different cultures. The next *anti-period drama* is equally extreme in its alienating beauty.
It was conceived by a British director and *metteur en scène*, a woman, passed master at
integrating televisual aesthetics into cinematic aesthetics, for an overwhelming effect.

Arnold-Hetreed (2011, UK, 129 min)

After an intense adaptative phase and a difficult shoot on the wind-blown moors of
North York, Andrea Arnold and Olivia Hetreed's *Wuthering Heights* came to life on the
big screen of the foreign Venice International Film Festival, on 6th Sept, 2011.[434] Their
prestigious *anti-period drama*, described five days later in *The Observer* as 'a wild child of
a film, a runaway, a cheeky git, a prickly hedgehog' by film critic Jason Solomons received
a wide coverage in the British media. A frenzy of activity would sweep in its wake hired
reviewers, inspired commentators and web-loggers alike. This feverish interest asserted
the vigour of the original literary text, and mirrored the earnestness of a film translation,
which not only reflects but also generates light, and integrates the eye of the director
of photography Robbie Ryan into the three-dimensional Figure of *Mise en Abîme* that
hallmarks Emily Brontë's Gothic tale.[435]

More than ever, the context as much as the content of the film translation needs to be
elucidated. The 2011 *Wuthering Heights* did not only create a sensation because of the
unstoppable move of the Arnold-Hetreed tandem towards more *racial authenticity* in the
representation of Heathcliff, but also because of Arnold-Ryan 'self-conscious decision
to test film boundaries and commercial sensibilities', which, as Solomons noticed, was

not 'a very British thing to do'. In the fervour and flaws of this 'un-British', pro-mythical period drama is ingrained an overwhelming desire to do justice to complex characters responding to their inhospitable environment, and to involve the spectators in their journeys. In the professional make-up of Arnold, a dancer, an actress, a screenwriter, a director and a *metteur en scène*, is patent a Free Cinema ethos resulting, for her *Wuthering Heights*, in the casting of youngsters, without any previous acting experience, in the main parts of teenage Cathy and Heathcliff (Shannon Beer and Solomon Glave, respectively) and of twenty-year-old Heathcliff (James Howson). This desire to do justice to complex characters and involve the spectators in their journeys characterised the other two quite 'un-British' film-makers who also made the pride of British reviewers at Venice that year, namely, the American director of Swedish and Japanese descent, Cary Joji Fukunaga, with his breath-taking *Jane Eyre* and the British director of Grenadian and Trinidian descent, Steve McQueen, with his controversial *Shame*. Together with Andrea Arnold's *Wuthering Heights*, their respective creations had contributed to what Solomons aptly described as a 'big [British] splash at the Lido'.[436] Interestingly enough, Arnold and McQueen both felt compelled to defend their well-considered cinematic choices to the press, on the release of their films on the UK screens:

> I wonder whether my bleak-o-meter is set differently from other people's. […] I have such passion for what I do that I can't see it as bleak. When people use that word, or 'grim' or 'gritty', I just think, 'Oh, come on, look a bit deeper'. My films don't give you an easy ride. I can see that. The sense I get is that people have quite a physical experience with them. They feel afterwards that they've really been through something.[437]
>
> (Arnold)

> I made *Shame* in America, but it's not a Hollywood movie. I'm about challenging people. Like, properly challenging them and their assumptions. Audiences make their minds up about people they see on screen, just like they do in real life. That's what fascinates me in film. You see a character and have to think: is this person different to what I assumed he was when I first saw him?[438]
>
> (McQueen)

In Arnold's *Wuthering Heights*, the complex character who occupies and, sometimes, even re-arranges the thoughts of the cinema viewers is the Black British Heathcliff. By means of a strict focalisation on the character of young teenager Heathcliff, the spectators experience with him the new territory of the family moor farm, and the exhilarating open space of the surrounding heathland, which he discovers through all his senses, and explores in great detail through the cinematographer's lens – or Robbie Ryan's eye. In the best takes, Ryan's lens signals its presence by diffracting the light, and gives the viewers the possibility to spot, in the 4:3 aspect ratio of television serials (which, according to Arnold, is 'the perfect frame for a person'), the Gothic Figure of the Intra-

Diegetic Observer Heathcliff inside its three-dimensional archetypal architecture.[439] On the morning following his first night at the Heights, young teenager Heathcliff stands by the parlour window seat where, in a later episode, little Cathy settles down to watch the rain-lashed heath, and, from this comfortable vantage point, catches a glimpse of her, barefoot, emptying her bucket in the nearby field. He focuses on her through the small glass panes of the mullioned window, fixes the images of her in his mind then, after a little pause, risks himself outdoors. While relishing the swirls of wind and every atom of space, he is hit hard by Hindley's galloping horse. Promptly rescued by Cathy, who leads him into the dusty stables where she deftly prepares their horse, he comes back to his senses and they end up riding together to Penistone Crags, the natural refuge high up on the moor that epitomises their childhood love. In a beautiful instantiation of hyper-subjective camera, since Heathcliff – or his elusive shadow, Ryan – seems to be carrying the eye of the camera within his own self all the time, the new kid on the moors registers all he perceives, the warmth at the flank of the horse, the texture of Cathy's hair, the very density of the air, and is rewarded by the rare heathen panorama visible only from the top of the Crags.

For a former enslaved child deprived of his most elementary rights, experiencing this strong scopic drive and these acute sensations is synonymous with experiencing freedom and, in his first examination of Cathy Earnshaw through the panes of the mullioned window, Andrea Arnold's Heathcliff is reminiscent of Emily Brontë's rich tenant Lockwood in his first clear apprehension of Catherine Heathcliff as she appears to him 'un-sheltered' from the shadows of Wuthering Heights, in Chap. II (Vol. I).[440] By fusing the narratorial trajectories of spectator and lover, and intertwining the Self-Revelatory Journey of the Intra-Diegetic Observer (and Invader) Lockwood with that of the Black Orphan Heathcliff, Arnold's translation is 'superior in depth of recapture' to Peter Bowker's (2009) that never *compensates* for the loss of Lockwood's scopic drive in the film narrative.[441]

Heathcliff's exclusivity as bearer of the camera, which works so well during the first hour of the Arnold movie as the young teenager Heathcliff adjusts to a favourable environment and a new language then, at the death of Earnshaw, endures proudly the degradation imposed by his White Brother Hindley, becomes problematic when Heathcliff returns to the Grange as a successful young man, under the traits of James Howson. This glitch in the story-retelling was widely noticed, and mainly ascribed to an increase in dialogue, and to a change of actors:

> With both roles now taken by older actors [James Howson as Heathcliff, Kaya Scodelario as Cathy], the movie never recovers its early power and at times becomes confused, ponderous, and risible.[442] (Philip French)

d more of a problem was the faint stiffness and self-consciousness of the the crucial lack of chemistry between the adult Heathcliff and Cathy. We :lieve in this love in order for Arnold's gloriously bruised and brooding visiʊ.. properly hit home and I never did, quite. This duo don't like us; they won't hold our gaze. So all we can do is sit in the dark and admire their travails from afar, like peering through binoculars at some big cat at play on open ground; one that is too wild – too unwilling – to draw too close.[443] (Xan Brooks)

Those reviewers' comments seem to point out to some form of individual aesthetic threshold that needs to be crossed in order to appreciate the Arnold movie. Its strong un-melodramatic fervour goes against the grain of Philip French's vibrant nostalgia for the 'monochrome classics of seven decades ago'; its overpowering flavour forces Laurence Olivier's Heathcliff down from his pedestal, transforming the classic monochrome character into 'a puzzle, a tornado of resentment whirling destructively across the bleak and intimidating landscape'.[444] Arnold's 'televisual' (or even 'transmedial') aesthetics translates into an unusual 'cinematic' lack in safety distance between spectators and actors: this makes it difficult for Xan Brooks to find his bearings as a sheltered participative spectator.

A flaw in the dynamic structures of story-retelling, however, must also account for the reviewers' mixed reception of the older actors' impersonations of Heathcliff (James Howson) and Cathy (Kaya Scodelario). To obtain a well-balanced translation, the complex, acclimatising character with whom the hyper-subjective camera should be attached, when Heathcliff returns to the Grange as a successful young man, is Cathy. Her estranging journey from the Heights to the Grange, which she undergoes physically in the absence of Heathcliff as he leaves and she gets married, but also prolongs mentally, in the very opposite direction, as he returns, corresponds to an enforced ellipsis inasmuch as the bias of the Intra-Diegetic Narrator, Ellen Dean (Simone Jackson), completely obliterates her perspective. Through an individual effort of their imagination, the readers of the novel might be better prepared (as Extra-Diegetic Observers) for the dormant narratorial storm that will be unleashed years after the original scission in the Earnshaw family and departure of Heathcliff, *as he returns*. The fragmentary recounting (by the controlling Ellen) of Cathy's first estranging illness, and of her consecutive move to Thrushcross Grange, fitted in the space of a few pages straddling Chap. IX and X (Vol. I),[445] not only unveils the psychological significance of Cathy's second estranging illness, which breaks out on the occasion of a vigorous fight between Heathcliff and Edgar in Chap. XI (Vol. I),[446] but also of Ellen's narratorial bias.

In the Arnold movie, switching the hyper-subjective camera from Heathcliff to Cathy, *as he returns transformed into another actor*, could have turned into a focal point this 'opaque centre' in Emily Brontë's novel where Cathy's perspective is momentarily lost. This entry on stage of the returning Heathcliff as James Howson also signals the entry on stage of Cathy Linton, the Lady of the Manor, interpreted by Kaya Scodelario. Under-cover of her grand title and amber adornments, Cathy is still painstakingly coming to terms with her break-up with Heathcliff and acclimatising to her new place in Thrushcross Grange.

This missed changeover in the gender and identity of the bearer of the camera, which might have balanced the translation and the viewpoints, was identified by Andrea Arnold herself, and related in those terms to journalist and film critic, Ryan Gilbey:

> It's such a complex book that I just had to pick out the things that had resonance to me, while still honouring the work as a whole. I knew I wanted to keep the kids in the film for the first hour, whereas most people only show them for ten minutes then move on to the adults. [...] I knew then that I couldn't squeeze everything in. The journey has taught me so much about what I feel towards the material. I was even thinking *I might have cast a woman as Heathcliff*. That would have been interesting. [...] I don't want to give too much away because part of that feeling is still in the film, and I want people to discover it for themselves.[447] [Emphasis added]

A switch mechanism triggering the brief apparition, on a nocturnal background of heath, of each white and phantomatic letter shared between the title of the *Wuthering Heights* novel and film, had established a bond between the subjectivity of the twenty-year-old Heathcliff who, at the end of the movie, in the immediate aftermath of Cathy's death, bashes his head against the walls of the old bedroom – still bearing the carved letters of her name – and that of the very young orphan Heathcliff making his way through some unfamiliar and coarse grasses, on the stormy night of his arrival at the isolated moor farm of the Heights, as the white letters of the movie's title are materialising on the cinema screen. This switch in hyper-subjectivity had connected the two different actors who incarnate Heathcliff at two liminal moments in time, the end and the beginning of the story as retold in the movie. It had also associated the Gothic Figure of *Mise en Abîme* created on screen by the emergence of the white letters on an inscrutable screen of moors and wilderness, with an emblematic passage in the novel involving the Intra-Diegetic Invader Lockwood:

> The ledge, where I placed my candle, had a few mildewed books piled up in one corner; and it was covered with writing scratched on the paint. This writing, however, was nothing but a name repeated in all kinds of characters, large and small – *Catherine Earnshaw*, here and there varied to *Catherine Heathcliff*, and then again to *Catherine Linton*.
>
> In vapid listlessness I leant my head against the window, and continued spelling over Catherine Earnshaw – Heathcliff – Linton, till my eyes closed; but they had not rested five minutes when a glare of white letters started from the dark, as vivid as spectres – the air swarmed with Catherines; and rousing myself to dispel the obtrusive name, I discovered my candle wick reclining on one of the antique volumes, and perfuming the place with an odour of roasted calf-skin.[448]

Considering Arnold and Hetreed's inventiveness in splitting and re-imagining this Fantastic episode for the screen, a switch in hyper-subjectivity with a changeover from Heathcliff to Cathy as the bearer of the camera, *as Heathcliff returns*, would not only have been workable but also represented a huge progression within the specific type of narration developed throughout the *mise en scène*. Realistically, however, as Arnold might also have half-confided to Gilbey, this exercise in perfectionism would mean working twice as much for achieving an equivalent presence of Cathy's subjectivity on screen. This could have irremediably delayed the actual making of a film that 'taught [Arnold] so much about what she felt towards the material', and was, in itself, a very practical exercise in humility:

> We all knew it would be a difficult location, but we weren't prepared for how tough it was. It was so muddy that your feet sank wherever you walked. We were shooting in a house that kept collecting water. We ate lunch huddled around this metal heater in a stinky old sheep-house. On the last week, *I was carrying a heavy camera box on my head*, walking up this hill, and my legs just gave way. I collapsed to my knees and burst into tears. But I didn't want anyone to see me crying, so I pulled my hat down over my eyes and sat there on the grass with the box on my head. Eventually I struggled to the top of the hill, where I saw another crew member, also *with a camera box on her head*, also on her knees crying. We'd had enough.[449] [Emphasis added]

With hindsight, and from the cosy perspective of a historical and transnational journey, the powerful mechanism of 'hyper-subjectivity swapping' suggested by Arnold's endeavours could finally have prevented an over-rationalisation of Cathy's early pregnancy. Her pregnancy, in the Arnold-Hetreed script, is exploited as a simple catalyser to the plot and, since the heroine never becomes the bearer of the camera, deprived of most of its potential impact as an Integrational Indice that could have reflected the dynamic structures of Cathy's psyche.[450] As in the Buñuel movie, Cathy's pregnancy dictates the smothering attitude of the household at Thrushcross Grange, and accounts for her refusal to follow Heathcliff (James Howson). The latter, having caught sight of the tiny piece of garment that Ellen, delighted, shows to Cathy, pleads that he would bring up the child 'as his own' without – as far as he can see – winning Cathy back, or altering the course of future events. From then on, in Heathcliff's mind, revenge against the past and the present dominates everything, but what is going on in Cathy's mind?

The more classic strategy of externalised camera at work in Cary Fukunaga's impressive *Jane Eyre*, and most of the other *Wuthering Heights* film translations, can prove to be quite revealing of the characters' psyches, especially when the film translation is treated as a pro-mythical drama, Gothic in essence, where the unidentified bearer of the camera is discreetly watching the characters, and de-multiplying the points of view. Because the UK release date of the Fukunaga *Jane Eyre* fell two months before the first shows of the Arnold *Wuthering Heights*, the Charlotte Brontë heroine, 'Miss Eyre', and the evocative cinematography and score translating her fusional encounter with the moors, came to validate my own imaginings of Cathy Earnshaw as an acclimatising character. Jane Eyre is at the centre of the spectators' gaze when she explores, candle in hand, the half-familiar, half-nightmarish Thornfield Hall, and so is Cathy Earnshaw when she unveils to Ellen the colliding landscapes of Thrushcross Grange, Wuthering Heights and the moors, and of all the threats gathering around her with her second illness.

The Cathy of Emily Brontë's novel whose identity is shattered by a relived estrangement, and a pregnancy undiscernable to her, is neither the Cathy of Hetcht and MacArthur, brought on to her deathbed by a paroxysm of jealousy, nor the Cathy of Bowker, drained of her vital forces by a mad hike to Penistone Crags as she is heavily pregnant. The Cathy of Emily Brontë's novel is the reverse side of the homeless Heathcliff (or orphaned Onimaru); she is the Jane Eyre of Fukunaga's film, lost at a desolate crossroads, but seeking actively her way on the moorland, and put back on track by the spectre of her childhood friend, Helen Burns.[451] The externalised camera that works around her, and catches the light bouncing off the grand wooden floors, the beds of rocks and the moisture-suffused scenery is, in Fukunaga's words, 'amplifying the expression of Jane's soul'. This externalised camera is also magnifying the sincerity of the acting in the midst of the most passionate dialogue scenes, and showcasing Jane's contained panic and anguish – simply 'displaced into the keening of a single violin'[452] – when she escapes from Rochester's grasp and struggles to find a light, on the dark ocean of the moors. Judging from Philip French, the aforementioned film critic who, in the Autumn of 2011, wrote in *The Observer* a critical piece on *Wuthering Heights* following his review of *Jane Eyre*, the Fukunaga adaptation was 'a good-looking film, serious, thought through and well acted' that was nonetheless lacking 'the cinematic intensity of the Orson Welles version'.[453]

Fukunaga's *Jane Eyre* that won the United Kingdom the Goya for Best European Film, and Mia Wasikowska the Best Actress Award from the Independent British Film Awards, could not (or would not) be fully embraced.[454] The un-melodramatic restraint of the acting, the cinematic distance that Adriano Goldman's externalised camera puts between spectators and actors, had somehow its faults, and Philip French's critical piece echoed the opinion of the most visible and influential figures in the British cultural press: neither of those two films, *Jane Eyre* or *Wuthering Heights*, were deemed to have found the right adaptative template.[455]

However, through the hyper-subjective camera borne by Heathcliff, the Cathy of Emily Brontë's novel translated, and quite convincingly too, into the Cathy Earnshaw of Andrea Arnold's film. At the Valladolid International Film Festival, shortly before its release in the United Kingdom, the younger interpreter of Cathy Earnshaw, thirteen-year-old Shannon Beer, was rewarded with the Younger Actors Prize. Her prize, as well as her young partner's prize, were confirming the impact of Andrea Arnold's *mise en scène*, and complementing Robbie Ryan's technical achievement, which the Golden Osella had celebrated in Venice, with a recognition based primarily on artistic performance.[456] Arnold who, interestingly, never refers in her interviews to her scriptwriter Olivia Hetreed – the adaptor of Tracy Chevalier's novel, *Girl with a Pearl Earring* – seems to have given the adaptive impulse and overall script orientation to the *Wuthering Heights* project. Nothing so far has transpired of the fine-tuning of the dynamic structures of the Imaginary that led to the selection of the Integrational Indices and Catalysers to the Plot, or of Arnold's share in the actual writing of the script itself. Arnold rooted her structural and aesthetical choices in the depths of the Gothic architecture of *Mise en Abîme* or, in other words, in the Realm of the Lovers' Journey. In the absence of Lockwood and his Substitutes, who are the Primary (or Intra-Diegetic) Outsiders, the journey of the Cinematic Spectators – the Secondary (or Extra-Diegetic) Outsiders – is left entirely dependent on their empathy with Heathcliff, and on their immersion into the Lovers' Journey. Therefore, the Cinematic Spectators become, by force, 'Outsiders': Outsiders, when they cannot adhere completely to Heathcliff's angle of vision and remain foreign to the reality of the film; Outsiders, when they follow Heathcliff in his liberating journey away from racism and social exclusion, and become 'Liberated Outsiders' themselves. Nowhere is Heathcliff more liberated than when he, the new kid on the moors, experiences Nature through Nicolas Becker's soundtrack 'incorporat[ing] wind, birdsong, barking dogs, rain, the flapping of shutters, the whispering of leaves, the chattering of insects and the creaking of trees', and becomes Cathy Earnshaw's alter ego through observation and language:

> The movie is at its strongest in these early scenes as Cathy and Heathcliff form a childhood bond against the bitter world and become one with each other in the natural world, suggestive of the friendship that young slaves briefly forged with their masters' children in the deep south [sic].[457]

Through the hyper-subjective camera borne by Arnold's Heathcliff, Emily Brontë's Cathy can be apprehended as the teacher on the moors, but differs from her daughter Catherine who teaches Hareton how to read with books imported from Thrushcross Grange.[458] Arnold's Cathy transmits Heathcliff her knowledge of the surrounding moors, and of the English idiom, without books, and her alter ego's learning is indistinguishable from their interplay with Nature. At the end of their first ride to Penistone Crags, they discover in a swift close-up the skeletons of some small mammals, then spot some simple feathers: these remains do not constitute an eerie sight, they are indicators of the time going by on the moorland. The bones belong to the silent Earth and the primordial alluvium; the feathers to the Air and the realm of Words. Heathcliff, who bared his teeth when the dogs assailed him on the night of his arrival at the Heights, is no more a feral child than Cathy is when she first meets him, and directly spits at him, the intruder. He speaks an African dialect that the inhabitants of Wuthering Heights cannot understand: when her father first brought him in, Cathy heard him using the foreign words against the dogs. She will transform, in the scene following their return from the Crags, the wooden floor of her room into a desk where all the feathers and quills collected on the hills are deftly arranged. Cathy is thus transmitting her familial and cultural lore by sharing with Heathcliff her knowledge of the birds that live on the moor, and by teaching him their names. She knows how to match any bird's feather with the correct English word – the lapwing's is her favourite – and, when she is older and Heathcliff visits her at Thrushcross Grange years later, she would be using a lapwing's feather as her page marker.

First, little Cathy shows Heathcliff the feather of the bird, then pronounces clearly its name, which he repeats willingly after her. They both enjoy this language game that is born of feathers swirling skywards in the wind, and is sealed by their shared experience of the deathly embrace in the bog. Without her love, without their non-verbal exchanges, he would not have taken part in the cultural game, and could not have blended the new indigenous language into his own.

Arnold's Cathy is a natural political activist, she is Heathcliff's liberator and changes him into an Insider who retains this intense scopic drive typical of the Outsider. She increases, in the process of liberating Heathcliff, her own faculties of observation, and becomes a liberated Insider – a potential bearer of the camera. Arnold's translation, like Yoshida's, bases 'all *emotional impact* on communicating the *sympathy* with [they feel] for the *transgressor*', 'The Enemy' of the *Mumford and Sons* song that accompanies the last sequence. In that last sequence, the deathly embrace in the bog is being re-played, and it will be prolonged into the end credits with 'The Enemy' as unique soundtrack.[459] This dissenting angle of vision brings on to the surface the deeply political vein in Emily Brontë's novel, the vein that Lady Eastlake feared when, in her article of December 1848, she insultingly likened Cathy and Heathcliff to 'the Jane and Rochester animals in their native state', implying at the time '*Irish* (or *vagrant* animals) in their native state'.[460] This is also the vein that David Sexton pillaged, in his review of November 2011, by comparing Howson's Heathcliff to 'a young Rio Ferdinand', even though he was probably well aware that Arnold had initially looked for her Heathcliff in the travellers' community that lives in the United Kingdom.[461]

In the aftermath of Arnold's beautiful yet alienating movie, British film critic Steve Rose and Black British actor Joseph Paterson, both reflected on the topicality of her cultural translation of *Wuthering Heights*. They did so in two equally invigorating articles, 'How Heathcliff got a *racelift*' and 'Why *Wuthering Heights* gives me hope', and shall be given the last, if not conclusive, words of our translators' journey, which is also a mythocritical journey. In assessing the degree of involvement (or 'foreignisation') of the observers-viewers, we have been actively seeking and revealing the Other:

Through Heathcliff, *Wuthering Heights* unsettled formerly stable boundaries of 19th-century Britain, including racial ones, and it is apparently still doing it today. You could see Heathcliff as the first post-racial hero – coping with a world that wasn't yet ready for him, and still isn't.

(Steve Rose)

Not only is it essential that we as British people tell our story, it is vital that we tell the whole story. If not, we risk increasing those feelings of alienation and temporariness that effect [sic] our youth so violently. […] As an actor, I want to be in works that reflect black presence in the UK throughout the nation's history. But if I am to do that, then playwrights must get researching to broaden their palette, and programme-makers must look away from their mirrors and see the darker shades around both them and their ancestors. In the meantime, I applaud Arnold's intelligence and openness in casting who she liked, regardless of their ethnicity.[462]

(Joseph Paterson)

Figure 20. Simplified Chart of the Dynamic Structures of *Wuthering Heights*

	Lockwood or His Substitute(s), the Intra-Diegetic Outsider(s)	Cathy-Heathcliff Catherine-Hareton
First Mythical Component (MC1)	Correspondence between Settings and Characters	Connection between the Profane and the Sacred through Nature and the Mind
Second Mythical Component (MC2)	Fantastic Visions and Liminal Places associated with Childhood	Oneiric Connection between the Living and the Dead; Correspondence between Childhood and the Sacred
Third Mythical Component (MC3)	Self-Revelatory Romantic Journey	Initiatory Love Journey; Correspondence between Love and the Sacred
First Bataillan Theme (BT1)	New and Hostile Territories; the Present Moment (Trespassing-Learning 1)	The Ephemeral Sovereignty of Childhood; the Present Moment and the Sacred (Interdict 1)
Second Bataillan Theme (BT2)	Personal Development versus Social Constraints; Empathy with Evil (Trespassing-Learning 2)	Hyper-Morality and Morality; Revolt of Evil against Good (Interdict 2)
Third Bataillan Theme (BT3)	Life-Changing Moral Transformation (Trespassing-Learning 3)	From Eros to Thanatos and Back (Interdict 3)
Planar Figure	Rejection and Identification	Opposites and Doubles
Gothic Figure	*Mise en Abîme*	Lineage

Acknowledgements

T here have been quite a few encounters fuelling this study, which started as an interview project based on my interest in a number of film and television adaptations of *Wuthering Heights*, most notably Jacques Rivette's *Hurlevent*, but also many other creations that were not 'classics' and sprang from very diverse generational and cultural quarters – Wainwright's *Sparkhouse*, Yoshida's *Onimaru* and Tilley and Fuest's *Wuthering Heights*, for instance. Because of my commitment to the Kino ArtHouse Cinema and associated Cork Film Festival, it quickly made sense for me to prolong my spectator's experience into an interactive experience that would shed light on both the industry-related and creative sides of the adaptive process.

Resorting each time to a personalised set of interview questions structured around the close-reading of each film or television adaptation, I set out to interact (whenever possible) with the practitioners who had gone through the daunting task of translating Emily Brontë's novel (and its unconscious) for the screen – or, in the case of producer Philippe Jacquier, had contributed to the production of Yoshida's *later* transcultural work – but who, without regretting their involvement one single minute, had not very much benefited from the risky endeavour. Their interest resided in the challenge of the adaptative process more than in fame or money, and my thanks have to go to them first.

This research work, which entailed a lot of transcription rewrites, editing and translation work, ended up forming an interesting collection of interviews, and caught the eye of the Centre for Film Studies, UCD, Ireland. As I started my Ph.D. on 'The Cultural Adaptations of *Wuthering Heights*', it soon became clear, though, that I could not weave a scholarly tapestry out of those seemingly disconnected interviews, reviews and budding critical essays without exploring 'archaeologically' the lost silent movie by Albert V. Bramble and Eliot Stannard. As I was to find no expertise at UCD for this crucial leg of the journey, this is really down to the intelligence and generosity of Charles Barr, who put me in touch with Ian Macdonald and Mary Hammond, if the precursor to *A Journey Across Time and Cultures* came to light at all.

From there, the vigour of the *Spaces of Television* blog (designed for the same-named AHRC-funded project) encouraged me to trust in the quality of the archival research I had originally carried out at the BBC Written Archives (BBC WA) in Reading, and evolve towards a progressive shedding of the archival artefacts themselves for some longer and more confident strands of analysis of the BBC Plays and *Classic Serials*. The attentive James Codd and the rest of the team at BBC WA contributed to keeping the panic of the referencing of the material at bay, when I worked on the final version of the manuscript. Further, their positive frame of mind inadvertently made up for the deception that was mine when Harumi Balthus Klossowski de Rola, after having suggested that it might be possible, eventually said 'no' to the incorporation of five of her father's previously published Persian-ink drawings into the Rivette section, on the grounds that giving them up to me for free would make her high-profile Copyright lawyer look redundant!

The *West Lothian Writers*, who supported 'Nine Months to Make a Book', the community-based project running parallel to the publishing of this monograph, as well as writers Aintzane Legarreta Mentxaka and Richard Pierce-Saunderson, all gave some extra momentum to the final leg of this journey into the creative depths of *Wuthering Heights*. Those colleagues and friends never tried to dampen my enthusiasm by reminding me that I was not writing in my native language, or pretend that they could not care about the transmission of fragments of an old myth to new audiences, since they are writers themselves. My thanks go to them heartily too.

Last but not least, Welsh academic Liz Jones managed to read between the lines of the original manuscript and, in doing so, prompted me to write out what I had quite simply omitted to reference clearly, 'la mythocritique', because it belonged to a category of very 'accented' academic writings primarily applied to literary texts, and seldom mentioned in British and Anglo-American Film, Television and Media Studies – or even in contemporary Adaptation Studies at large.

The support of the production team at Intellect never gave way. I salute the patience of my Production Manager Claire Organ. On a more personal level, I salute the endurance of Philip whose hospitality and friendship have been crucial over the years and, of course, the steadfastness and love of Mark, my husband.

Along the way, Louis and Armand grew up and learnt how to read, write stories, play football, and sing Scottish and Austrian songs. They enjoy watching 'foreign' fictions on the small family television set, as much as going out to the Edinburgh Filmhouse to catch animation movies and warm-hearting classics on the large cinema screens. Thank you for never being bored and having no intention whatsoever of grasping the concept of mono-culturalism.

Postscript

Wuthering Heights on Film and Television: A Journey Across Time and Cultures (1920-2011) is as much typified by its recognition of the work of some fine television and film historians, anthropologists of the Imaginary, and Adaptation Studies specialists as by its dedication to showcase the inventive work of those who stand in the front line of **adaptation** seen as a **mode of transmission and foreignisation**, the 'adaptators' themselves. In this light, I would like the '*Black & White Artist*' Albert V. Bramble and his lost *Wuthering Heights* (1920) to remain forever present in your mind. Through archival research and mythocritical associations, he resurrects from the erasure of casual critical discourse and breaks new ground in British film history.

Bramble overwhelmingly represents the Other, the Black Orphan, the Ghost-Child. These figures of Outsiders are at the heart of most Fantastic urban tales, which also use, in some fashion or another, the Gothic architecture of *Mise en Abîme* to bring the reluctant postmodern spectators in, transform their greyish surroundings into exotic ones and revive some old fragments of disinterested love. As a contributor to Adaptation Studies and writer of short stories constructed for the ubiquitous space of the small screen, I believe it would be great for the world of Arts & Humanities to open up much more to translingual figurative structuralism and to integrate fully into their prose those writings interspersed with foreign, non-strictly academic voices that belong to film and television critics, biographers, interpreters-philosophers, translators, writers, producers and actors.

Bibliography and Filmography

I. Film Aesthetics and Critical Studies

French References

Daney Serge and Deleuze Gilles, *Ciné journal,* Vols. 1 & 2 (Paris: Petite bibliothèque des *Cahiers du Cinéma*, 1988).

Le Roux Hervé, 'Un pas en arrière, deux pas en avance', *Cahiers du Cinéma*, no. 371-372 (May 1985), pp. 25-27.

Sato Tadao, *Le cinéma japonais*, Tome II (Lavaur, France: cinéma/pluriel, Centre Georges Pompidou, 1997).

Tessier Max, *Le cinéma japonais: Une introduction* (Saint-Germain-du-Puy, France: Collection 128, Nathan Université, 1997).

English References

Berg Charles M., 'The Human Voice and the Silent Cinema', *Journal of Popular Film and Television*, Vol. 4, no. 2 (1975), pp. 165-177.

Boorman John and Donohoe Walter, *Projections 2: A Forum for Film Makers* (London: Faber & Faber, 1993).

Bordwell David, 'Poetics of Cinema', in David Bordwell, *Poetics of Cinema* (New York and London: Routledge, 2007).

Bordwell David, 'Common Sense + Film Theory = Common-Sense Film Theory?' (David Bordwell's website on Cinema: Essays, May 2011).

Brewster Ben and Jacobs Lea, *Theatre to Cinema: Stage Pictorialism and the Early Feature Film* (Oxford: Oxford University Press, 1997).

Brownlow Kevin, 'Silent Films: What was the Right Speed?', *Sight and Sound*, Vol. 49, no. 3 (1980), pp. 164-167.

Brunel Adrian, *Filmcraft* (London: George Newnes, 1933).

Bruzzi Stella, *Undressing Cinema: Clothing and Identity in the Movies* (London and New York: Routledge, 1997).

Burch Noël, *Life to Those Shadows/La Lucarne de l'Infini* (Berkeley: University of California Press, 1990).

Burch Noël, *To the Distant Observer: Form and Meaning in Japanese Cinema* (Berkeley: University of California Press, 1979).

Gledhill Christine, *Home is Where the Heart is: Studies in Melodrama and the Woman's Film* (London: BFI Books, 1987).

Gledhill Christine, *Melodrama and Realism in 1920s British Cinema*, Pamphlet (London: BFI Editions, 1991).

Gledhill Christine, 'Coda: Hitchcock, *The Manxman* and the Poetics of British Cinema', in *Reframing British Cinema (1918-1928): Between Restraint and Passion* (London: BFI Publishing, 2003), pp. 119-122.

Gunning Tom, 'From the Opium Den to the Theatre of Morality: Moral Discourse and the Film Process in Early American Cinema' (Chap. 8), in Lee Grieveson and Peter Krämer (eds), *The Silent Cinema Reader* (London and New York: Routledge, 2004).

Gunning Tom, 'The Scene of Speaking: Two Decades of Discovering the Film Lecturer', *Iris*, no. 27 (1999), pp. 67-80.

Higson Andrew, *Waving the Flag: Constructing a National Cinema in Britain* (Oxford: Oxford University Press, 1995).

Higson Andrew (ed.), *Young and Innocent? The Cinema in Britain 1896-1930* (Exeter: University of Exeter Press, 2002).

Lindsay Vachel, *The Art of the Moving Picture*, 1915 (New York: The Macmillan Co., 1922).

Low Rachael, *The History of the British Film 1919-1929* (London: George Allen & Unwin, 1971).

Low Rachael, *The History of the British Film 1929-1939: Film Making in 1930s Britain* (London: George Allen & Unwin, 1985).

Macdonald Ian W., 'Struggle for the Silents: The British Screenwriter from 1910 to 1930', *Journal of Media Practice*, Vol. 8, no. 2 (September 2007), pp. 115-128.

Macdonald Ian W., 'The Silent Screenwriter: The Re-discovered Scripts of Eliot Stannard', *Comparative Critical Studies (British Comparative Literature Association)*, Vol. 6, no. 3 (2009), pp. 385-400.

Macdonald Ian W., *The Poetics of Screenwriting* (Basingstoke: Palgrave Macmillan, 2013).

McFarlane Brian (ed.), *The Cinema of Britain and Ireland*, preface by Roy Ward Baker (London: Wallflower Press, 24 Frames, 2005).

McKee Robert, *Story, Substance, Structure, Style and the Principles of Screenwriting* (New York: Regan Books/Harper Collins, 1997).

Metz Christian, *The Imaginary Signifier/Le Significant imaginaire: Psychoanalysis and the Cinema*, trans. Celia Britton, Annwyl Williams and Ben Brewster (Bloomington and Indianapolis: Indiana University Press, 1982).

Seldes Gilbert, *The Seven Lively Arts: The Classic Appraisal of the Popular Arts,* 1924 (Mineola, New York: Dover Publications, 2001).

Stannard Eliot, 'Writing Screen Plays', Lesson Six in the series *Cinema – in Ten Complete Lessons* (London: Standard Art Book Company, 1920).

Thompson Kristin, 'The Formulation of the Classical Style, 1909-28', in David Bordwell, Janet Staiger and Kristin Thompson (eds), *The Classical Hollywood Cinema: Film Style & Mode of Production to 1960* (New York: Columbia University Press, 1986).

Turvey Gerald, 'Enter the Intellectuals: Eliot Stannard, Harold Weston and the Discourse on Cinema and Art', in Alan Burton and Laraine Porter (eds), *Scene-stealing: Sources for British Cinema before 1930* (Trowbridge: Flicks Books, 2003), pp. 85-93.

II. Critical Writings on Television

Barr Charles, '*They Think It's All Over*: The Dramatic Legacy of Live Television', in John Hill and Martin McLoone (eds), *Big Picture, Small Screen: The Relations Between Film and Television*, Acamedia Research Monograph 16 (Luton: University of Luton Press, 1986), pp. 47-75.

Bignell Jonathan, 'The Spaces of The Wednesday Play (BBC TV 1964–1970): Production, Technology and Style', 'Spaces of Television' Special Issue of the *Historical Journal of Film, Radio and Television* (11th Sept, 2014), pp. 369-389.

Cartier Rudolph, 'A Foot in Both Camps', *Films and Filming*, 4.12 (September 1958), pp. 10 and 31.

Giddings Robert and Selby Keith, *The Classic Serial on Television and Radio* (Basingstoke: Palgrave Macmillan, May 2001).

Jacobs Jason, *The Intimate Screen: Early British Television Drama* (Oxford: Clarendon Press, 2000).

Kneale Nigel, 'Not Quite So Intimate', *Sight and Sound*, Vol. 28 (1959), pp. 86-88.

McMahon Kate, 'The Broadcast Interview: Laura Mackie, ITV Drama', *Broadcast Now* (25th Feb, 2009).

Porter Michael J., Larson Deborah L., Harthcock Allison and Berg Nellis Kelly, 'Redefining Narrative Events: Examining Television Narrative Structure', *Journal of Popular Film and Television*, Vol. 30, no. 1 (Spring 2002), pp. 23-30.

Renoir Jean and Rossellini Roberto, interview with André Bazin, 'Cinema and Television', *Sight and Sound,* Vol. 28, no. 1 (Winter 1958–1959), pp. 26-27.

Silvey Robert, *Who's Listening: The Story of BBC Audience Research* (London: Allen & Unwin, 1974).

Sutton Shaun, *The Largest Theatre in the World: Thirty Years of Television Drama* (London: BBC Books, 1982).

III. Critical Writings on Film Adaptation and Translation

French References

Bellemin-Noël Jean, *Gradiva au pied de la lettre. Relecture du roman de W. Jensen dans une nouvelle traduction* (Paris: puf, Coll. *Le Fil rouge*,1983).

Berman Antoine, *L'épreuve de l'étranger. Culture et traduction dans l'Allemagne romantique: Herder, Goethe, Schlegel, Novalis, Humboldt, Schleiermarcher, Hölderlin* (Paris: Gallimard Essais, 1984).

Berman Antoine, *L'âge de la traduction. 'La tâche du traducteur' de Walter Benjamin, un commentaire* (Vincennes: Presses universitaires de Vincennes, 2008).

Cocteau Jean, *La Belle et la Bête: Journal d'un film* (Paris: J.B. Janin, 1946).

Ostria Vincent, 'Du papier au celluloid', *Cahiers du Cinéma*, Vol. 371-372 (May 1985), pp. 124-128.

Piaget Jean, *La Formation du symbole chez l'enfant* (Neuchâtel and Paris: Delachaux et Nieslé, 1945).

Piaget Jean, *Introduction à l'épistémologie génétique* in 3 Vols. (Paris: puf, 1950).

English References

Aragay Mireia, *Books in Motion: Adaptation, Intertextuality, Authorship* (Amsterdam and New York: Rodopi, 2005).

Benjamin Walter, 'The Task of the Translator', in *Illuminations* Harry Zohn (trans.) & Hannah Arendt (intro. and ed.) 1923 (New York: Harcourt Brace Jovanovich, 1968).

Benjamin Walter, *The Work of Art in the Age of Its Technological Reproducibility, and Other Writings on Media*, 1936 (Cambridge (MA) and London (UK): Harvard University Press, 2008).

Berman Antoine, *The Experience of the Foreign: Culture and Translation in Romantic Germany*, trans. Stefan Heyvaert (Albany: SUNY Press, 1992 [1984]).

Bernstein Matthew Leonard, 'In Light of the Aura: Benjamin's Aesthetics in Contemporary Fiction', BA thesis (Middletown: Wesleyan University, 2011).

Cartmell Deborah, Hunter I.Q. and Kaye Heidi (eds), *Pulping Fictions: Consuming Culture Across the Literature/Media Divide* (London and Chicago: Pluto Press, 1996).

Cartmell Deborah and Whelehan Imelda (eds), *Adaptations: From Text to Screen, Screen to Text* (London and New York: Routledge, 1999).

Cattrysse Patrick, 'Film (adaptation) as Translation: Some Methodological Proposals', *Target*, Vol. 4, no. 1 (1992), pp. 53-70.

Cattrysse Patrick, 'The Unbearable Lightness of Being: Film Adaptation Seen from a Different Perspective', *Literature/Film Quarterly*, Vol. 25, no. 3 (1997), pp. 222-230.

Cattrysse Patrick, 'Stories Travelling Across Nations and Cultures', *Meta: journal des traducteurs/Meta: Translators' Journal*, Vol. 49, no. 1 (2004), pp. 39-51.

Cattrysse Patrick, 'Film Adaptation as Translation: the Polysystem approach revisited', *Journal of Adaptation in Film and Performance* (2011).

Cattrysse Patrick, *Descriptive Adaptation Studies: Epistemological and Methodological Issues* (Antwerp and Apeldoorn: Garant Publishers, April 2014).

Chamberlain Lori, 'Gender and the Metaphorics of Translation', in Lawrence Venuti (ed.), *Rethinking Translation: Discourse, Subjectivity, Ideology* (London: Routledge, 1992), pp. 57-74.

Chatman Seymour, *Coming to Terms: The Rhetoric of Narrative in Fiction and Film* (Ithaca: Cornell University Press, 1990).

Cobb Shelley, 'Adaptation, Fidelity, and Gendered Discourses', in *Adaptation*, Vol. 4, no. 1 (2011), pp. 28-37.

Cobb Shelley, *Adaptation, Authorship, and Contemporary Women Filmmakers* (London: Palgrave Macmillan, 2014).

Dudley Andrew, 'The Well-Worn Muse: Adaptation in Film History and Theory', in Syndy Conger and Janice R. Welsch (eds), *Narrative Strategies* (Macomb: West Illinois University Press, 1980).

Dudley Andrew, *Concepts in Film Theory* (New York: Oxford University Press, 1984).

Eisenstein Sergei, 'Dickens, Griffith and the Film Today', in *Essays in Film Theory: Film Form*, 1944, edited and translated by Jay Leyda in 1977 (San Diego, New York & London: A Harvest Book, Harcourt, 1990s), pp. 195-255.

Elliott Kamilla, 'Literary Cinema in the Form/Content Debate' (Chap. 5), in *Rethinking the Novel/Film Debate* (Cambridge: Cambridge University Press, 2003), pp. 133-183.

Giddings Robert and Sheen Erica (eds), *The Classic Novel: From Page to Screen*, Chap. II (Manchester: Manchester University Press, 2000), pp. 14-30.

Gouanvic Jean-Marc, 'A Bourdieusian Theory of Translation, or the Coincidence of Practical Instances: Field, Habitus, Capital and Illusion', 'The Translator' Special Issue of *Bourdieu and the Sociology of Translation and Interpreting*, Vol. 11, no. 2 (2005), pp. 147-166.

Hammond Mary, 'Hitchcock and *The Manxman*: A Victorian Bestseller on the Silent Screen' (University of Southampton, 2008), in R. Barton Palmer (ed.), *Hitchcock at the Source* (New York: SUNY, 2011).

Jakobson Roman, 'On Linguistic Aspects of Translation', 1959, in Rainer Schulte and John Biguenet (eds), *Theories of Translation* (Chicago: University of Chicago Press, 1992), pp. 144-151.

Klein Michael and Parker Gillian (eds), *The English Novel and the Movies* (New York: Frederick Ungar, 1981).

Kranz David L. and Mellerski Nancy C., *In/Fidelity: Essays on Film Adaptation* (Newcastle, UK: Cambridge Scholars Publishing, 2008).

Krebs Katja, *Translation and Adaptation in Theatre and Film* (London and New York: Routledge, 2014).

Leitch Thomas, 'Between Adaptation and Allusion', in *Film Adaptation and Its Discontents: From Gone With the Wind to The Passion of Christ* (Baltimore and London: Johns Hopkins University Press, 2009), pp. 93-126.

Leitch Thomas, 'Vampire Adaptation', *Journal of Adaptation in Film and Performance*, Vol. 4, no. 1 (2011), pp. 5-16.

Macedo Ana Gabriela and Esteves Pereira Margarida, 'Identity and Cultural Translation – Writing Across the Borders of Englishness: Women's Writing in English in a European Context', in *European Connections* (Bern, Switzerland: Peter Lang, 2005).

McFarlane Brian, 'It wasn't Like That in the Book', *Literature/Film Quarterly*, Vol. 28, no. 3 (2000), pp. 163-169.

McFarlane Brian, 'A Literary Cinema? British Films and British Novels', in Charles Barr (ed.), *All Our Yesterdays: 90 Years of British Cinema* (London: BFI 1986), pp. 120-142.

McKechnie Kara, 'Gloriana – the Queen's Two Selves: Agency, Context and Adaptation Studies', in Monika Pietrkak-Franger and Eckart Voigts-Virchow (eds), *Adaptations: Performing Across Media and Genres* (Trèves, Germany: Wissenschaftlicher Verlag Trier, 2009), pp. 193-209.

Minier Márta, 'Reconsidering Translation From the Vantage Point of Gender Studies', in *Identity and Cultural Translation – Writing Across the Borders of Englishness: Women's Writing in English in a European Context* (Bern, Switzerland: Peter Lang, 2005), pp. 59-83.

Minier Márta, 'Definitions, Dyads, Triads and Other Points of Connection in Translation and Adaptation Discourse', in Part I (Converging Agendas) of *Translation and Adaptation in Theatre and Film* (London and New York: Routledge, 2014), pp. 13-35.

Murray Simone, 'Books as Media: The Adaptation Industry', *The International Journal of the Book*, Vol. 4. no. 2 (2007), pp. 23-30.

Naremore James (ed.), 'Introduction: Film and the Reign of Adaptation', in James Naremore (ed.), *Film Adaptation* (London: Athlone Press, 2000), pp. 1-16.

Raw Laurence, 'Bridging the Translation/Adaptation Divide: A Pedagogical View', in Part III (Emerging Practices) of *Translation and Adaptation in Theatre and Film* (London and New York: Routledge, 2014), pp. 162-177.

Raw Laurence, 'Identifying Common Ground', in *Translation, Adaptation and Transformation* (London, New Delhi, New York and Sydney: Bloomsbury, 2012).

Robinson Douglas, *The Translator's Turn* (Baltimore and London: Johns Hopkins University Press, 1991).

Robinson Douglas, *Displacement and the Somatics of Postcolonial Culture* (Columbus: Ohio State University Press, 2013).

Sanders Julie, *Adaptation and Appropriation* (London and New York: Routledge, The New Critical Idiom Series, 2006).

Sheen Erica, 'Introduction', in R. Giddings and E. Sheen (eds), *The Classic Novel: From Page to Screen* (Manchester: Manchester University Press, 2000), pp. 1-13.

Stam Robert, 'Beyond Fidelity: The Dialogics of Adaptation', in James Naremore (ed.), *Film Adaptation* (London: Athlone, 2000), pp. 54-76.

Stam Robert, 'Introduction: The Theory and Practice of Adaptation', in R. Stam and A. Raengo (eds), *Literature and Film: A Guide to the Theory and Practice of Film Adaptation* (Oxford: Blackwell, 2005), pp. 1-52.

Steiner George, *After Babel: Aspects of Language and Translation*, 1st ed. 1975 & 2nd ed. 1992 (Oxford: Oxford University Press, 1998, 3rd ed.).

Steiner George, 'The Hermeneutic Motion', in Lawrence Venuti (ed.), *The Translation Studies Reader* (New York and London: Routledge, 2000).

Stoneman Patsy, *Brontë Transformations: The Cultural Dissemination of Jane Eyre and Wuthering Heights*, Chap. IV: The Inter-war Period, 'The World's Great Love Stories': Brontës recuperated, 1920-1944 (Hemel Hempstead: Prentice Hall/Harvester Wheatsheaf, 1996), pp. 114-116.

Van Parys Thomas, Review article of David L. Kranz and Nancy C. Mellerski's *In/Fidelity: Essays on Film Adaptation*, *Image & Narrative*, Vol. 14, no. 1 (2013), pp. 148-153.

Venuti Lawrence, *The Translator's Invisibility: A History of Translation*, 1995 (Abingdon, Oxon, UK: Routledge, 2008, 2nd ed.).

Venuti Lawrence, 'Adaptation, Translation, Critique', *Journal of Visual Culture*, Vol. 6, no. 1 (2007), pp. 25-43.

Vincendeau Ginette, *Film/Literature/Heritage: A Sight and Sound Reader* (London: BFI, 2001).

Woolf Virginia, 'The Cinema', in L. Woolf (ed.), *Collected Essays*, 1926, Vol. 2 (London: Chatto and Windus).

IV. Critical Writings on the Film Adaptations of *Wuthering Heights*

French References

Baecque Antoine de, '*Onimaru*: Les dieux et les hommes', *Cahiers du Cinéma*, no. 413 (1988), pp. 52-53.

Bergala Alain, '*Onimaru*: Combustion froide', *Cahiers du Cinéma*, no. 409 (1988), p. 14.

Berthomieu P., 'From *Wuthering Heights* to *The Big Country*: William Wyler's Approach to Filming Indoor Settings and Wide Open Spaces', *Positif*, no. 484 (June 2001), pp. 83-85.

Chevrie Marc, 'La main du fantôme', *Cahiers du Cinéma*, no. 376 (October 1985), pp. 4-7.

Guibert Hervé, '*Hurlevent*: Entretien avec le cinéaste Jacques Rivette', *Le Monde* (10th Oct, 1985).

English References

Attia Nadia, 'Curzon Interview: Andrea Arnold (ONLINE EXCLUSIVE)', *Curzon Cinemas Online Magazine* (Thursday 29th Sept, 2011).

Bluestone George, '*Wuthering Heights*', in *Novels into Film: The Metamorphosis of Fiction into Cinema*, 1957, Chap. III (Berkeley: University of California Press, 1961).

Bowden Bryony, '*Wuthering Heights*: Peter Kosminsky', *Times Literary Supplement* (October 1992), p. 17.

Bradshaw Peter, '*Jane Eyre* – Review', *The Guardian* (Thursday 8th Sept, 2011).

Brooks Xan, '*Wuthering Heights* – Review', *The Guardian* (Tuesday 6th Sept, 2011).

Brooks Xan, 'Andrea Arnold Finds New Depths in *Wuthering Heights*', *The Guardian* (Tuesday 6th Sept, 2011).

Brooks Xan, 'Dubbed: The Actors Who Lost Their Voices', *The Guardian* (Monday 14th Nov, 2011).

Canby Vincent, 'Buñuel's Brontë', *The New York Times* (27th Dec, 1983).

Catania Saviour, 'Wagnerizing *Wuthering Heights*: Buñuel's Tristan storm in *Abismos de pasión*', *Literature/Film Quarterly* (1st Oct, 2008).

Christensen Inger, 'Too Unreal: The Representation of Catherine in Wyler's *Wuthering Heights*', *Literary Women on the Screen*, Chap. I. 3 (Bern, Switzerland: Peter Lang Inc., 1991), pp. 32-53.

Clarke Cath, 'Kaya Scodelario Scales New *Wuthering Heights*', *The Guardian* (Thursday 8th Sept, 2011).

Conrad Peter, 'Jane Eyre and Wuthering Heights: Do We Need New Film Versions?', *The Observer* (Sunday 21st Aug, 2011).

Crace John, 'Wuthering Heights, Atlantic Convoys', *The Guardian* (Monday 31st Aug, 2009).

Crist Judith, 'Wuthering Depths', *New York Magazine* (22nd Feb, 1971), p. 68.

Dawtrey Adam, 'Twilight Gives New Brontë Films Wings', *The Guardian* (Wednesday 2nd Dec, 2009).

Fragola Anthony N., 'Buñuel's Re-vision of *Wuthering Heights*: The Triumph of *L'amour fou* over Hollywood Romanticism', *Literature/Film Quarterly*, Vol. 22, no. 1 (1994), pp. 50-56.

Francke Lizzie, '*Wuthering Heights*: Peter Kosminsky', *Sight and Sound*, Vol. 2, no. 6 (October 1992).

French Philip, '*Jane Eyre* – Review', *The Observer* (Sunday 11th Sept, 2011).

French Philip, '*Wuthering Heights* – Review', *The Observer* (Sunday 13th Nov, 2011).

Gassner John and Nichols Dudley (eds), '*Wuthering Heights*', in *Twenty Best Film Plays*, 1943 (New York: Crown Publishers, 1959), pp. 293-331.

Gilbey Ryan, 'Andrea Arnold: I don't do Easy Rides', *The Guardian* (Monday 31st Oct, 2011).

Harrington John, 'Wyler as Auteur', in Michael Klein and Gillian Parker (eds), *The English Novel and the Movies* (New York: Frederick Ungar, 1981).

Jeffries Stuart, 'Sex and Rebellion: *Desperate Romantics* Writer Peter Bowker on His New BBC Drama', *The Guardian* (Tuesday 21st July, 2009).

Holmwood Leigh, 'TV Ratings: *Framed* Beats *Wuthering Heights* in Bank Holiday Drama Battle', *The Guardian* (Tuesday 1st Sept, 2009).

Jones Julie, 'Fatal Attraction: Buñuel's Romance with *Wuthering Heights*', *Anales de la Literatura Española Contemporanea*, Vol. 22, no. 1-2 (1997), pp. 149-163.

Kneale Nigel, 'An Electrifying Masterpiece', *Radio Times* (4th Dec, 1953), p. 14.

Martin Amy, 'A Battle on Two Fronts: Wuthering Heights and Adapting the Adaptation', in Part I (Adapt) Chap. 3 of *Film Remakes, Adaptations and Fan Productions: Remake/ Remodel* (London and New York: Palgrave Macmillan, 2012), pp. 67-86.

Martin Sara, 'What does Heathcliff Look Like? Performance in Peter Kosminsky's Version of Emily Brontë's *Wuthering Heights*', in *Books in Motion* (Amsterdam and New York: Rodopi, 2005). pp. 52-67.

Mayer Sophie, 'The New Wuthering Heights does not Ignore Racism; It tackles It Full On', *The Guardian* (Thursday 8th Dec, 2011).

Miller Lucasta and Rose Cynthia, 'How Cathy Came Home', *New Stateman & Society*, 16th Oct, 1992, pp. 33-34.

Mills Pamela, 'Wyler's Version of Brontë's storms in *Wuthering Heights*', *Literature/Film Quarterly*, Vol. 24, no. 4 (1996), pp. 414-422.

Paterson Joseph, 'Why *Wuthering Heights* Gives Me Hope', *The Guardian* (Friday 11th Nov, 2011).

Plunkett John, 'Sparkhouse Proves to be a Damp Squid', *The Guardian* (2nd Sept, 2002).

Popkin Michael, '*Wuthering Heights* and Its Spirit', *Literature/Film Quarterly*, Vol. 15, no. 2 (1987), pp. 116-122.

Rose Steve, 'How Heathcliff Got a Racelift', *The Guardian* (Sunday 13th Nov, 2011).

Russell Williams Imogen, 'How the Brontës Divide Humanity', *The Guardian* (Thursday 23rd Sept, 2010).

Schmidt Rick, '*Wuthering Heights*', *Variety* (16th Dec, 1970).

Secher Benjamin, 'Dark depths of Andrea Arnold's *Wuthering Heights*', *The Telegraph* (Saturday 5th Nov, 2011).

Sexton David, '*Wuthering Heights* – Review', *The London Evening Standard* (Friday 11th Nov, 2011).

Solomons Jason, 'Venice Film Festival: Britain's Big Splash at the Lido', *The Observer* (Sunday 11th Sept, 2011).

Spence Martin, 'The Attempts to Film the Life and Works of the Brontë Sisters: Green Slime and Devotion', *Sight and Sound*, Vol. 56 (1987), pp. 215-217.

Tilley Patrick, '*Wuthering Heights*', *The Times* (18th Apr, 1970).

Tomalin Claire, 'Height of Romance', *Radio Times* (23rd Sept, 1978), pp. 5-7.

Wall Ian, *Emily Brontë's Wuthering Heights* (London: Film Education Study Guide, 1992).

Walsh John, *Wuthering Heights*, Andrea Arnold, 128 mins, 15; *The Rum Diary*, Bruce Robinson, 120 mins, 15, in *The Independent* (Sunday 13th Nov, 2011).

V. Emily Brontë: Useful Works and Biographies

French Reference

Brontë Emily Jane, *Cahiers de poèmes*, traduit de l'anglais et présenté par Claire Malroux and Pierre Leyris (Paris: José Corti, 1995).

English References

Allott Miriam (ed.), *The Brontës: The Critical Heritage* 1974 (London and Boston: Routledge & Kegan Paul, 1995).

Barnett David, 'How Much Harm does a Bad Book Cover Do?', *The Guardian* (Wednesday 19th Aug, 2009).

Brontë Emily, *Wuthering Heights*, 1847, Introduction and Notes by Pauline Nestor, Preface by Lucasta Miller (London: Penguin Books, 2003).

Cecil David (Lord), 'Emily Brontë and Wuthering Heights' (Chap. 5), in *Early Victorian Novelists: Essays in Revaluation*, 1934 (Indianapolis: Bobbs-Merrill, 1935).

Daiches David, 'Foreword' to *Wuthering Heights*, 1965 (Harmondsworth: Penguin English Library Edition, 1979) pp. 7-29.

Gérin Winifred, *Emily Brontë: A Biography* (Oxford: Clarendon, 1971).

Qi Shouhua and Padgett Jacqueline, *The Brontë Sisters in Other Wor(l)ds* (New York and London: Palgrave Macmillan, October 2014).

Sanger Charles Percy, *The Structure of Wuthering Heights* (London: Hogarth Essays XIX, Hogarth Press, 1926).

Stoneman Patsy, 'Emily Brontë – Wuthering Heights', in *Readers' Guides to Essential Criticism* (London and New York: Palgrave Macmillan, 2000).

Winnifrith Tom and Chitham Edward, *Charlotte and Emily Brontë: Literary Lives* (Basingstoke, Hampshire: Macmillan, 1989).

VI. Biographies, Interviews and Essays about Film-Makers and Television Programme-Makers

French and Spanish References

Bazin André, 'William Wyler ou le janséniste de la mise en scène', *La revue du cinéma*, nos. 10 & 11 (February and March 1948), pp. 46-47 & pp. 58-60, respectively.

Ciment Michel and Viviani Christian (ed.), 'Dossier William Wyler', *Positif*, no. 484 (June 2001), pp. 68-102.

Daney Serge and Narboni Jean, 'Entretien avec Jacques Rivette', *Cahiers du Cinéma*, no. 327 (September 1981), pp. 8-21.

Doniol-Valcroze Jacques and Bazin André, 'Entretien avec Luis Bunuel', *Cahiers du Cinéma*, no. 36 (June 1954), pp. 2-14.

Drouzy Maurice, *Luis Buñuel, architecte du rêve* (Paris: Série La Mise en Film/Éditions Pierre Lherminier, 1978).

Gester Julien, 'Yoshida, trésor caché', *Les inrockuptibles*, no. 644 (2008), pp. 48-49.

Jacquier Philippe and Pranal Marion, *Gabriel Veyre, Opérateur Lumière* (Arles, France: Institut Lumière/Actes Sud, 1996).

Monegal Antonio, *Luis Buñuel: De la Literatura al Cine, Una Poetica del Objeto* (Barcelona: Anthropos, 1993).

Morain Jean-Baptiste and Lalanne Jean-Marc, 'Entretien Jacques Rivette: L'art secret', *Les inrockuptibles*, no. 590 (2007), pp. 30-35.

Niogret Hubert, 'Interview with Kiju Yoshida', *Positif*, no. 506 (April 2003), pp. 8-11.

Païni Dominique, 'Jacques Rivette: Les finalités incertaines', *Cahiers du Cinéma*, no. 376 (October 1985), p. 50.

Pinon-Kawatake Josiane, 'Interview with Yoshida Kijû (Yoshishige)', *Positif*, no. 312 (1987), pp. 23-26.

English References

Aranda José Francisco, *Luis Buñuel: A Critical Biography*, 1969 (New York: Da Capo Press, 1976).

Barr Charles, 'Hitchcock's British films revisited', in Andrew Higson (ed.), *Dissolving Views: Key Writings on British Cinema* (Cassell: London, 1996), pp. 2-19.

Barr Charles, 'Hitchcock and Early Filmmakers', in Thomas Leitch and Leland Poague (eds), *A Companion to Hitchcock* (Wiley-Blackwell Companions to Film Directors, 2011), pp. 48-66.

Barr Charles, *English Hitchcock* (Moffat, Scotland: Cameron and Hollis, 1999), pp. 27-57.

Barr Charles, 'Writing Screen Plays: Stannard and Hitchcock', in Andrew Higson (ed.), *Young and Innocent?* (Exeter: The University of Exeter Press, 2002), pp. 227-241.

Berg A. Scott, *Goldwyn: A Biography* (New York: Alfred A. Knopf, 1989).

Buñuel Luis, *My Last Breath*, 1982 (New York: Alfred A. Knopf, 1983).

Buñuel Luis, de la Colina José and Pérez Turrent Tomás, *Objects of Desire: Conversations with Luis Buñuel* (*Prohibido asomarse al interior*, 1986), translated by Paul Lenti (Marsilio: New York, 1992), p. 160.

Callan Michael Feeney, *Richard Harris: Sex, Death and the Movies – An Intimate Biography* (London, UK: Robson, 2003), pp. 146-158.

Creely Niamh, 'Think Tank', *Film Ireland*, no. 130 (1ˢᵗ Sept, 2009).

Eaton Michael, 'The Man Who wasn't There', *Sight and Sound, Script special: Eliot Stannard*, Vol. 15, no. 12 (December 2005), pp. 18-19.

Elsaesser Thomas, 'Rivette and the End of Cinema', *Sight and Sound,* Vol. 1, no. 12 (April 1992), pp. 20-23.

Evans Peter William, *The Films of Luis Buñuel: Subjectivity and Desire* (New York: Oxford University Press, 1995).

Fiddy Dick, 'Peter Hammond: Actor Who Became a Prolific TV Director', *The Guardian* (Sunday 1ˢᵗ Jan, 2012).

Gassner John and Nichols Dudley, 'Wuthering Heights', in *Twenty Best Film Plays* (New York: Crown Publishers, 1943), pp. 294-331.

Gaughan Gavin, 'Robert Fuest: Film Director Who Worked with Vincent Price', *The Independent* (4ᵗʰ Nov, 2011).

Gaughan Gavin, 'Peter Hammond: Stage and Screen Actor Who went on to Direct Classic Television Shows', *The Independent* (25ᵗʰ Apr, 2012).

Gilbey Ryan, 'Andrea Arnold: I don't do Easy Rides', *The Guardian* (Monday 31ˢᵗ Oct, 2011).

Hepworth Cecil, *Came the Dawn* (London: Phoenix House, 1951).

Herman Jan, *A Talent for Trouble: William Wyler*, translated and edited by David Robinson, 1995 (New York: Da Capo Press, 1997).

Kerzoncuf Alain and Barr Charles, *Hitchcock: Lost and Found* (Lexington, Kentucky: University of Kentucky Press, 2015).

Magee Karl, 'Hooray for Hollywood? The Unmade Films of Lindsay Anderson' (Chap. 7), in Dan North (ed.), *Sights Unseen: Unfinished British Films* (Cambridge: Cambridge Scholars Publishing, 2008), pp. 121-140.

Marshall Bill, *André Téchiné* (Manchester: Manchester University Press, French Film Directors Series, 2007).

Myles Lynda and Petley Julian, 'Rudolph Cartier', *Sight and Sound*, Vol. 59-60 (1990), pp. 126-129.

Pérez Turrent Tomas, *Luis Buñuel in Mexico* (London: *British Film Institute*, 1995), pp. 116-120.

Powell Michael, *A Life in Movies: An Autobiography* (New York: Knopf, 1987).

Preston Adam, 'I was Buñuel's double', *Sight and Sound, Script Special: Jean-Claude Carrière*, Vol. 15, no. 12 (December 2005), pp. 20-21.

Solomons Jason, 'Steve McQueen: I Could Never Make American Movies – They like Happy Endings', *The Observer* (Sunday 8ᵗʰ Jan, 2012).

Truffaut François, *Hitchcock* (New York: Simon & Schuster, 1967).

Wyler William, 'No Magic Wand', *Screen Writer* (February 1947), pp. 22-23.

Yoshida Kiju, *Ozu's Anti-Cinema* (Ann Harbor: Center for Japanese Studies, University of Michigan, 2003).

Yoshida Kiju, 'My Theory of Film: A Logic of Self-Negation', *Review of Japanese Culture and Society*, Vol. 22 (December 2010), pp. 104-109.

VII. Literary, Historical, Anthropological, Narratological and Mythocritical Works

French References

Alain-Fournier Henri, *Le Grand Meaulnes*, 1913 (Paris: Le livre de poche, 1990).

Bachelard Gaston, *L'Air et les songes* (Paris: José Corti,1943).

Bachelard Gaston, *La Terre et les rêveries du repos* (Paris: José Corti, 1948).

Bachelard Gaston, *La Terre et les rêveries de la volonté* (Paris: José Corti, 1947).

Barthes Roland, *Le Degré zero de l'écriture* (Paris: Éditions du Seuil, 1953).

Barthes Roland, *Mythologies* (Paris: Éditions du Seuil, 1957).

Barthes Roland, *S/Z essai sur Sarrasine d'Honoré de Balzac* (Paris: Éditions du Seuil, 1970).

Barthes Roland, *L'Empire des signes* (Paris: Skira, 1970).

Barthes Roland, *Fragments d'un discours amoureux* (Paris: Éditions du Seuil, 1977).

Bataille Georges, *La littérature et le mal* (Paris: Gallimard, 1957).

Bellemin-Noël Jean, *Vers l'inconscient du texte* (Paris: puf, 1979).

Caillois Roger, *Au Coeur du fantastique* (Paris: Gallimard, 1965).

Chauvin Danièle, Siganos André and Walter Philippe, *Questions de mythocritique. Dictionnaire* (Paris: Éditions Imago, 2005).

Coustillas P., Petit Jean-Pierre and Raimond Jean, *Le roman anglais au XIXᵉ siècle*, Chap. IV (Paris: puf, 1978), p. 163–178.

Delecroix Michel, Rosselin Mary, *La Grande-Bretagne au XIXᵉ siècle: Technologies et modes de vies* (Paris: Masson, 1991).

De Rougemont Denis, *L'Amour et l'Occident* (Paris: Éditions Plon, Coll. *Présence*, 1939).

Durand Gilbert, *Les structures anthropologiques de l'Imaginaire*, 1960 (Paris: DUNOD, 11ᵉ edition, 1992).

Durand Gilbert, *L'Imagination symbolique*, 1964 (Paris: Quadrige/puf, 1993).

Durand Gilbert, 'Les chats, les rats et les structuralistes', Chap. III, in *Figures mythiques et visages de l'œuvre. De la mythocritique à la mythanalyse*, 1979 (Paris: DUNOD, 1992), pp. 86-120.

Eliade Mircea, *Mythes, rêves et mystères*, 1957 (Paris: Folio/Essais, Gallimard, 1996).

Fierobe Claude, *De Melmoth à Dracula* (Rennes: Éditions Terre de Brume, 2000).

Genette Gérard, *Figures III* (Paris: Éditions du Seuil, Coll. *Poétique*, 1972).

Genette Gérard, *Palimpsestes: La Littérature au second degré* (Paris: Éditions du Seuil, Coll. *Poétique*, 1982).

Lévi-Strauss Claude, *Tristes Tropiques* (Paris: Éditions Plon, 1955).

Lévi-Strauss Claude, *Anthropologie Structurale* (Paris: Éditions Plon, 1958).

Lévi-Strauss Claude, *Mythologiques*, t. IV: *L'Homme nu* (Paris: Éditions Plon, 1971).

Lévy Maurice, *Le roman gothique anglais 1764–1824*, Chap. X, 1968 (Paris: Albin Michel, 1995).

Lupasco Stéphane, *Logique et Contradiction* (Paris: puf, 1947).

Lupasco Stéphane, *Les Trois Matières* (Paris: Julliard, 1960).

Maupassant Guy de, *Le Horla et autres contes cruels et fantastiques*, 1883-1887 (Paris: Garnier frères, Coll. *Classiques Garnier*, 1989).

Minkowska Françoise, *De Van Gogh et Seurat aux dessins d'enfants*, in the Exhibition Catalogue (Paris: Musée pédagogique, 1949).

Petit Jean-Pierre, *L'oeuvre d'Emily Brontë: La vision et les thèmes* (Lyon: L'Hermès, 1977).

Raimond Jean, *La littérature anglaise*, 1986 (Paris: Que sais-je ?/puf, 1990).

Rialland Ivanne, 'La mythocritique en questions', *Acta fabula*, Vol. 6, no. 1 (Printemps 2005), URL: http://www.fabula.org/revue/document817.php, page consultée le 28 décembre, 2014.

Tournier Michel, *Vendredi ou les Limbes du Pacifique*, 1967 (Paris: *Éditions* Gallimard, Coll. *Blanche,* 1967).

Tournier Michel, *Le Vent Paraclet*, 1977 (Paris: Gallimard/*Folio*, 1990).

English References

Abrams M.H., *Natural Supernaturalism: Tradition and Revolution in Romantic Literature* (London: Oxford University Press, 1971).

Barchi Panek Melissa, *The Postmodern Mythology of Michel Tournier* (Newcastle-upon-Tyne: Cambridge Scholars Publishing, 2012).

Barthes Roland, 'The Death of the Author', in Lodge David (ed.), *Modern Criticism and Theory*, 1968 (London and New York: Longman, 1988), pp. 167-172.

Barthes Roland, *A Lover's Discourse: Fragments* (New York: Hill & Wang, 1978).

Bataille Georges, 'Emily Brontë', *in Literature and Evil*, 1957 (London: Calder & Boyars, 1973), pp. 3-17.

Bettelheim Bruno, *The Uses of Enchantment: The Meaning and Importance of Fairy Tales*, 1976 (New York: Alfred A. Knopf, 1986).

Burke Edmund, *A Philosophical Inquiry into the Origin of Our Ideas on the Sublime and Beautiful*, 1757 (New York: Harper & Brothers, 1844).

Campbell Joseph, *The Hero with a Thousand Faces* (New York: Pantheon Books, The Bollingen Series, 1949), pp. 36-38.

Campbell Joseph, *The Masks of a God: Creative Mythology* (New York: Viking Press, 1968), pp. 64-75.

Cecil David, *Early Victorian Novelists: Essays in Revaluation*, 1934 (London: Constable, 1960), p. 167.

Coleridge Samuel T., *Biographia Literaria*, 1817 (Oxford: Clarendon Press, 1907).

Dawson Terence, 'The Struggle for Deliverance from the Father: The Structural Principle of *Wuthering Heights*', *Modern Language Review* (1989), pp. 289-304.

Dawson Terence, 'An Oppression Past Explaining: The Structures of *Wuthering Heights*', in *Orbis Litterarum*, Vol. 44, no. 1 (1989), pp. 48-68.

Debroise Olivier, de Sa Rego Stella, *Mexican Suite: A History of Photography in Mexico* (Austin: University of Texas Press, 2001), pp. 141-142.

De Rougemont Denis, *Passion and Society* (London: Faber and Faber Limited, 1939).

De Rougemont Denis, *Love in the Western World*, 1940 (New York: A Fawcett Premier Book, 1966).

Eliade Mircea, *Myths, Dreams and Mysteries* (London and Glasgow: Fontana Library, 1968).

Eliot George, *Mr. Gilfil's Love Story*, in David Lodge (ed.), *Scenes of Clerical Life*, 1858 (Harmondsworth: Penguin, 1987) pp. 117-244.

Everitt Alistair, *An Anthology of Criticism* (London: Frank Cass & Co Ltd, 1967).

Frye Northrop, *Anatomy of Criticism: Four Essays*, 1957 (Princeton: Princeton University Press, 1990).

Gilbert Sandra and Gubar Susan, *The Madwoman in the Attic: The Woman Writer and Nineteenth Century Literature*, Chap. VIII (New Haven: Yale University Press, 1979).

Hazette Valérie V., '*Wuthering Heights:* An Exploration into a Many-Layered Fiction', MA thesis (Reims, France: Université de Reims, 1998).

Jung Carl Gustav, *Symbols of Transformation* (a revision of *Psychology of the Unconscious*, 1912), Bollingen Series, Vol. 5 (Princeton: Princeton University Press, 1967).

Jung Carl Gustav, *Psychological Types* (1921), Bollingen Series, Vol. 6 (Princeton: Princeton University Press, 1971).

Kristeva Julia and Clément C., *The Feminine and the Sacred*, translated by J.M. Todd (New York: Columbia University Press, 2001).

Lacan Jacques, *The Four Fundamental Concepts of Psychoanalysis*, 1973 (London: Peregrine/Penguin Books, 1987).

Leavis Q.D., 'A Fresh Approach to *Wuthering Heights*', in *Collected Essays: The Englishness of the English Novel*, Vol. 1, 1969 (Cambridge: Cambridge University Press, 1983), pp. 228-274.

Le Fanu Joseph Sheridan, *In a Glass Darkly*, 1872 (London: Wordsworth Classics, 1995).

Lévi-Strauss Claude, *Myth and Meaning* (London: Routledge & Kegan Paul, 1978).

Lewis Matthew G., *The Monk*, 1796 (Oxford: Oxford University Press, The World's Classics, 1980).

Maugham William Somerset, 'W. Somerset Maugham Presents *Wuthering Heights*', in *The Ten Greatest Novels Selected by W. Somerset Maugham* (Philadelphia, Toronto: The John C. Winston Company, 1948).

Moody T.W. and Martin F.X., *The Course of Irish History*, 1967 (Cork and Dublin, Ireland: Mercier Press, 1994).

Mooneyham White Laura, 'Melodramatic Transformation: George Eliot and the Refashioning of *Mansfield Park*', in *Persuasions: The Jane Austen Journal* (1st Jan, 2003).

Poe Edgar Allan, *Tales of the Grotesque and Arabesque*, 1840 (Whitefish, Montana: Kessinger Publishing, 2004).

Propp Vladimir Aioakovlevich, *Morphology of the Folktale*, 1928 (Austin: University of Texas Press, 1968), pp. 25-81.

Shelley Mary, *Frankenstein*, 1818 (London: Wordsworth Classics, 1966).

Smith Roch C., *Gaston Bachelard* (Boston: Twayne, 1982).

Smith Sheila., 'At Once Strong and Eerie: The Supernatural in *Wuthering Heights* and Its Debt to the Traditional Ballad', *Review of English Studies*, Vol. 43, no. 172 (November 1992), pp. 498-517.

Stewart Michael, *Melodrama in Contemporary Film and Television* (London and New York: Palgrave Macmillan, 2014).

Stoker Bram, *Dracula*, 1897 (London: Wordsworth Classics, 1997).

Twitchell J.B., 'The Living Dead', Chap. IV, in *The Vampire in Prose* (Durham: Duke University Press, 1981), pp. 103-142.

Van Ghent Dorothy, 'On *Wuthering Heights*', in *The English Novel Form and Function Language Review* (New York: Harper Torchbooks, 1953), pp. 153-170.

Williams Anne, 'Natural Supernaturalism in *Wuthering Heights*', *Studies in Philology*, Vol. 82, no. 1 (Winter 1985), pp. 104-127.

Williams Anne, 'The Child is Mother of the Man', *Cahiers Victoriens et Edouardiens*, Vol. 34 (October 1991), pp. 81-94.

Wordsworth William and Coleridge S.T., *Lyrical Ballads*, 1798 (London: Duckworth and Co., 1907).

VIII. Filmography

Cinema

Wuthering Heights (Bramble A.V., Ideal Film Company, UK, 1920).

Wuthering Heights (Wyler William, The Samuel Goldwyn Company, US, 1939).

Abismos de pasión (*Cumbres Borrascosas*)/*Wuthering Heights* (Buñuel Luis, Producciones Tepeyac, Mexico, 1953).

Wuthering Heights (Fuest Robert, American International Pictures, UK/US, 1970).

Hurlevent/*Wuthering Heights* (Rivette Jacques, La Cecilia/Renn Productions/Ministère de la Culture, France, 1985).

Arashi ga oka/Onimaru (Yoshida Kiju, Sépia Production, Japan, 1988).
Wuthering Heights (Kosminsky Peter, Paramount British Pictures, UK/US, 1992).
Wuthering Heights (Arnold Andrea, HanWay Film/Ecosse Films/Film4, 2011).

Television

Wuthering Heights (O'Ferrall More George and Davison John, BBC Unrecorded, UK, 1948).
Wuthering Heights (Cartier Rudolph and Kneale Nigel, BBC, UK, 1953 & remake 1962).
Wuthering Heights (Sasdy Peter and Leonard Hugh, BBC Archives, 4 episodes, UK, 1967).
Wuthering Heights (Hammond Peter and Snodin David, BBC Archives, 5 episodes, UK, 1978).
Wuthering Heights (Skynner David, London Weekend Television/WGBH/ITV, UK, 1998).
Sparkhouse (Wainwright Sally, Red Production Company/BBC1, UK 2002).
Wuthering Heights (Bowker Peter, Mammoth Screen/WGBH/ITV, UK, 2009).

IX. Useful Websites

http://blogs.reading.ac.uk/spaces-of-television/
http://bufvc.ac.uk/
crlc.paris-sorbonne.fr/FR/Page_programme_2013.php
http://cri.u-grenoble3.fr/
http://filmstudiesforfree.blogspot.co.uk/
http://sensesofcinema.com
http://thelesserfeat.blogspot.co.uk
http://www.bfi.org.uk
http://www.davidbordwell.net/blog/
http://www.fabula.org
http://www.rouge.com.au
http://www.sensesofcinema.com
http://www.screenonline.org.uk

Endnotes

Summary

1. The newer versions are the Andrea Arnold/Olivia Hetreed's cinema movie of Autumn 2011, and the Coky Giedroyc/Peter Bowker two-part television movie shown on ITV1 at the end of Summer 2009. Both films are entitled *Wuthering Heights*.

2.. 'La structure des mythes' (Chap. XI) of 'Magie et religion' in *Anthropologie structurale* (1958), p. 233 for the original French version. This reference is also available in its English translation on page 17 of Melissa Barchi Panek's *The Postmodern Mythology of Michel Tournier*:

    Myth, like the rest of language, is made up of constituent units. These constituent units presuppose the constituent units present in language when analyzed on other levels – namely phonemes, morphemes and sememes [...] they belong to a higher and more complex order. For this reason we shall call them *gross constituent units*.

Part I

3. Extracted from Jonathan Bignell's Introduction to *An Introduction to Television Studies* (2008), p. 1.

4. The etymology reads '*muthos*', the Greek for 'myth' or 'Sacred tale' ('hieros logos').

5. Melissa Barchi Panek's *The Postmodern Mythology of Michel Tournier*, p. 2.

6. Michel Tournier's *Le Vent Paraclet*, p. 179.

7. This title has also been translated into *The Wanderer or The End of Youth*, *The Lost Estate* and, much more rarely, *The Mysterious Domain*.

8. Claude Fierobe's *De Melmoth à Dracula* (2000), p. 27.
9. 'Of the Sublime', Part I, Section VII, p. 51, in Edmund Burke's *A Philosophical Inquiry into the Origin of Our Ideas on the Sublime and Beautiful* (1757).
10. *Early Victorian Novelists: Essays in Revaluation*, 1934 (London: 1960, Constable), p. 167.
11. Tzvetan Todorov's *Introduction à la littérature fantastique* (Paris: Seuil, 1970), Chap. 3, pp. 46-62. In 1973, it was translated as *The Fantastic: A Structural Approach to a Literary Genre*.
12. My translation from Roger Caillois's *Au Coeur du fantastique* (Paris: Gallimard, 1965), p. 161.
13. Under the heading, 'Specific Continuous Forms (Prose Fiction)', pp. 303-307 of the 1990 Penguin Books edition (London).
14. These decades saw Henry Grattan's political dream come true with the creation of an independent parliament (1782-1783). Unfortunately, they were also marked by the failed rebellion of Theobald Wolfe Tone and the *Society of United Irishmen* against the British occupier (1798) and the re-institution of the *Legislative Union* (1800), which put an end to the independent parliament.
15. Sheila Smith's essay 'At Once Strong and Eerie: The Supernatural in *Wuthering Heights* and Its Debt to the Traditional Ballad' in *Review of English Studies*, Vol. 43, no. 172 (November 1992), pp. 506-514.
16. Gilbert Durand's *Les structures anthropologiques de l'Imaginaire* (1960), p. 65.
17. Four Factory Acts were passed between 1818 and 1848. The 1847 Factory Act was coupled to the Ten Hours Act for which the 'Radical Tory', Richard Oastler, had ardently campaigned.

    The *People's Charter* was presented to the UK Parliament in 1838, 1842 and 1848 successively. It contained six points that summarised the Chartists' political demands. They wanted electoral constituencies founded on an equal number of voters, universal suffrage (for men only), a salary for the Members of Parliament, voting secrecy, annual elections and no more suffrage on the basis of money qualification.
18. *Vida de Lazarillo de Tormes* (1554) created by an anonymous author was followed by Charles Sorel's *Vraie histoire comique de Francion* (1622), Alain-René Lesage's *Histoire de Gil Blas* (1715-1735), Tobias Smollett's *Roderick Random* (1748) and finally Henry Fielding's *Joseph Andrews* (1742) and *Tom Jones* (1749).
19. Nellie Dean tells Lockwood that Heathcliff was found by Mr Earnshaw, the old master, on the streets of Liverpool, 'a tale of his [Earnshaw's] seeing it starving, and houseless, and as good as dumb in the streets of Liverpool where he picked it up and inquired for its owner'.
    This is quoted from Chapter IV, page 37 of the 2003 Penguin Classics edition. All the subsequent page citations will refer to that edition.

20. This dark Byronic hero seems to have emerged as much from Byron's own public persona as from *Childe Harold's Pilgrimage, Cantos* I & II (1812) *closely* followed by his *Oriental Tales*: from *The Giaour* (1813) to *The Corsair* (1814) and *Parisina* (1816).

21. Emily Brontë's *Wuthering Heights*, 1847 (London: Penguin Books, 2003), Vol. I, Chap. I, p. 5.

22. This comment was incorporated to her review of Charlotte's *Jane Eyre*, at a date when the identities of Currer Bell (Charlotte), Ellis Bell (Emily) and Acton Bell (Anne) had not been fully revealed or differentiated.

    Mrs Rigby also believed that Wuthering Heights contained more than 'Byronic subversiveness', detecting in it 'the same strain of thought that had contributed to a second revolution in France and led to near-insurrection by the Chartists in England'.

23. K.M. Petyt's *Emily Brontë and the Haworth Dialect: A Study of the Dialect Speech in 'Wuthering Heights'* (Keighley: A Yorkshire Dialect Society Publication, 1970), p. 45.

24. Bill Marshall, *André Téchiné* (Manchester: MUP, French Film Directors Series, 2007), Chap. 1 'Emergence', pp. 23-24.

25. 'A Fresh Approach to *Wuthering Heights*' (1968-1969), in *Collected Essays: The Englishness of the English Novel*, Vol. 1 (1969), pp. 228-274.

26. 'Height of Romance', *Radio Times*, 23rd Sept, 1978, pp. 5-7.

27. 'The Child is Mother of the Man' (1991), p. 87.

28. In 1953, Henri-Pierre Roché published, *Jules et Jim*, and, in 1956, the year François Truffaut met him, *Les deux anglaises et le continent*.

29. 'Green Slime and Devotion', *Sight and Sound* (1987), pp. 216-217.

30. *Beauty and the Beast* was given a novelistic form by Jeanne-Marie Leprince de Beaumont, in 1748, and re-created in the movie by Jean Cocteau, in 1946.

31. Anne Williams' 'The Child is Mother of the Man' (1991), pp. 91-92.

32. Michael Popkin's '*Wuthering Heights* and Its Spirit' in *Literature/Film Quarterly* (1987), p. 118.

33. Mircea Eliade's 'Preface to the Original Edition of *Myths, Dreams and Mysteries*' (1956), p. 17.

34. George Steiner's *After Babel* (1992), Chap. 6 'Topologies of Culture', p. 452.

35. See the very first sentence of *Love in the Western World*, Book I 'The Tristan Myth', Chap. 2 ' The Myth', p. 18.

36. De Rougemont, op. cit., pp. 19-20.

37. Anne Williams' 'Natural Supernaturalism in *Wuthering Heights*' in *Studies in Philology* (Winter 1985), p. 126.

38. See Matthew Bernstein's 'In Light of the Aura: Benjamin's Aesthetics in Contemporary Fiction' (2011), p. 6:

    The process of auratic perception necessitates complex perceptual oscillations, between the physical and the metaphysical, between sense perception and a natural immanence. At the very least, Benjamin's first formulation registers the important fact that aura is never exclusively a function of artistic production, that it emerges in and through an active awareness. To rephrase a quote from the philosopher Novalis, one of Benjamin's favorites – aura is an attentiveness.

39   'Occasion of the Lyrical Ballads, and the Objects Originally Proposed' in Samuel Taylor Coleridge's *Biographia Literaria* (1817), Chap. XIV.
40.   *Wuthering Heights*, Vol. I, Chap. IX, p. 81.
41.   *Wuthering Heights*, Vol. II, Chap. XIX, pp. 323-324, and then Vol. II, Chap. XX, p. 331.
42.   Anne Williams, op. cit., p. 120.
43.   *Wuthering Heights*, Vol. I, Chap. III, p. 25.
44.   *Wuthering Heights*, Vol. I, Chap. XI, pp. 108-109.
45.   Georges Bataille's 'Emily Brontë' in *La littérature et le mal/Literature and Evil* (1957), p. 16:

The world of Wuthering Heights is **the world of a hostile sovereignty**.

46.   Bataille, op. cit., pp. 5-6.
47.   Bataille, op. cit., pp. 9-10.
48.   Bataille, op. cit., p. 11.
49.   Bataille, op. cit., p. 4.
50.   Bataille, loc. cit.
51.   George More O'Ferrall (director) and John Davison (writer)'s *Wuthering Heights* (1948); Rudolph Cartier (director) and Nigel Kneale (writer)'s *Wuthering Heights* (1953 and remake 1962).
52.   George Bluestone's '*Wuthering Heights*' (Chap. III) in his *Novels into Film* (1957), pp. 99-100.
53.   The third essay of Northrop Frye's *Anatomy of Criticism* (1957) is entitled 'Theory of Myths' or 'Archetypal Criticism'.
54.   See Dudley Andrew's *Concepts in Film Theory* (1984), pp. 98-99.
55.   The whole of Julie Sanders' Part 2 in her monograph, *Adaptation and Appropriation* (2006), is dedicated to 'Literary Archetypes'. In particular, her Chap. 4 is dedicated to 'Myth and Metamorphosis' (pp. 63–81) and prolonged into Chap. 5 with 'Fairy Tale and Folklore' (pp. 82-94).
56.   See David L. Kranz and Nancy C. Mellerski's Introduction to *In/Fidelity: Essays on Film Adaptation* (2008), p. 5.
57.   Kranz and Mellerski, op. cit., p. 207.
58.   De Rougemont, op. cit., Book II 'The Religious Origins', Chap. 6 'Courtly Love: Troubadours and Cathars', pp. 78-86.

59. Márta Minier's 'Definitions, Dyads, Triads and Other Points of Connection in Translation and Adaptation Discourse' in Part I (Converging Agendas) of *Translation and Adaptation in Theatre and Film* (2014), p. 28.

60. See Thomas Van Parys' Review article (2013) of *In/Fidelity*, pp. 151-152.

61. Kara McKechnie's 'Gloriana – the Queen's Two Selves: Agency, Context and Adaptation Studies' in *Adaptations: Performing Across Media and Genres* (2009), p. 193.

62. See 10.2.5 'Fidelity as an Operational Norm' (p. 306) as part of Chap. 10 'Adaptation as a Teleological Process' in Patrick Cattrysse's DAS (2014).

63. James Naremore in his introduction to *Film Adaptation* (2000) entitled 'Film and the Reign of Adaptation', p. 8.

64. Minier, op. cit., p. 19.

65. p. 24.

66. Sanders, op. cit., p. 19.

67. p. 7 of the French edition.

68. p. 8 of the French edition and p. 4 of the American-English edition:

> *The figure is outlined (like a sign) and memorable (like an image or a tale).*

69. See *Figures mythiques et visages de l'oeuvre* (1979), p. 89.

70. See Cattrysse, op. cit., p. 30, under the heading 2.2 'Gap between Theory and Practice' in Part I 'PS Adaptation Studies':

> Following the post-positivist view on knowledge (William M.K. Trochim, 2006), most researchers would now agree that a (more or less) coherent set of research programs should be preferred because it would allow researchers to triangulate across methods and reveal different aspects of reality. Since no approach can ever be complete, *a multiple perspective approach*, as opposed to a single perspective approach, may reveal a less incomplete picture of reality. [Emphasis added]

71. Patrick Cattrysse's 'Stories Travelling Across Nations and Cultures' in *Meta: journal des traducteurs/Meta: Translators' Journal* (2004).

72. Cattrysse's DAS, Introduction, p. 17.

73. Barchi Panek, op. cit., pp. 17–18 ('Myth is Language' in Chap. One 'Into the Shadows').

74. Lévi-Strauss, op. cit., p. 255.

75. Gilbert Durand's *Les structures anthropologiques de l'Imaginaire* (1960), p. 63.

76. http://cri.u-grenoble3.fr

77. http://www.crlc.paris-sorbonne.fr/FR/Page_programme_2013.php

78. See p. 190 of Jean Bellemin-Noël's *Vers l'inconscient du texte* (1979):

> La mise en récit littéraire d'une production inconsciente comporte [...] les éléments nécessaires à son interprétation, à la reconstitution d'un discours de désir, sans référence ni à ce que l'on sait par ailleurs de l'auteur, ni à ce que raconte ses autres oeuvres, ni à l'idiosyncrasie débridée d'un lecteur.

79. SAI, p. 38.
80. In this subsidiary clause defining the *anthropological trajectory* ('*genèse réciproque* qui oscille du geste pulsionnel à l'environnement matériel et social, et vice versa'/'*reciprocal genesis* that alternates between the drive-motivated gesture and the material and social environment, and vice versa'), Durand insists on the ontological simultaneity of (and bijective influence between) the physio- and psycho-logical imperatives ('les impératifs bio-psychiques') and the demands of the milieu ('les intimations du milieu'). SAI, p. 38 and top of p. 39.
81. SAI, 'Tableau de la classification isotopique des images', pp. 506-507.
82. See pp. 8-9 of Laurence Raw's 'Identifying Common Ground' in his *Translation, Adaptation and Transformation* (2012).
83. See p. 89 of *Figures mythiques et visages de l'œuvre. De la mythocritique à la mythanalyse* (1979):

> [...] une structure n'est pas, n'a jamais été, cette forme statique et vidée volontairement de sens qu'un certain structuralisme admet seule à la dignité de structure

84. *Wuthering Heights*, Vol. I, Chap. XI, pp. 108-109.
85. *Wuthering Heights*, Vol. I, Chap. III, p. 22:

> my companion is impatient and proposes that we should appropriate the dairy woman's cloak, and have a scamper on the moors, under its shelter.

86. *Wuthering Heights*, Vol. II, Chap. XV, p. 288.

> I'll tell you what I did yesterday! I got the sexton, who was digging Linton's grave, to remove the earth off her coffin lid, and I opened it. I thought, once, I would have stayed there, when I saw her face again – it is hers yet – he had hard work to stir me; but he said it would change, if the air blew on it, and so I struck one side of the coffin loose – and covered it up – not Linton's side, damn it! I wish he'd been soldered in lead – and I bribed the sexton to pull it away, when I'm laid there, and slide mine out too. I'll have it made so, and then, by the time Linton gets to us, he'll not know which is which!

87.   **Lockwood's nightmare**, *Wuthering Heights*, Vol. I, Chap. III, pp. 19-26:

pp. 19-20 The ledge, where I placed my candle, had a few mildewed books piled up in one corner; and it was covered with writing scratched on the paint. (…)

In vapid listlessness I leant my head against the window, (…) I discovered my candle-wick reclining on one of the antique volumes, and perfuming the place with an odour of roasted calf-skin.

I snuffed it off, and very ill at ease, under the influence of cold and lingering nausea, sat up, and spread open the injured tome on my knee. (…)

p. 22 I began to nod drowsily over the dim page; my eye wandered from manuscript to print. I saw a red ornamented title … 'Seven Times Seven, and the First of the Seventy First. A Pious Discourse delivered by the Reverend James Branderham, in the Chapel of Gimmerden Sough.' (…) I sank back in bed, and fell asleep.

Alas, for the effects of bad tea and bad temper! What else could it be that made me pass such a terrible night? I don't remember another that I can at all compare with it since I was capable of suffering.

p. 23 (…) Oh, how weary I grew. How I writhed, and yawned, and nodded, and revived! How I pinched and pricked myself, and rubbed my eyes, and stood up, and sat down again, and nudged Joseph to inform me if he would ever have done!

p. 24 (…) And what was it that had suggested the tremendous tumult, what had played Jabes' part in the row? Merely, the branch of a fir-tree that touched my lattice, as the blast wailed by, and rattled its dry cones against the panes!

I listened doubtingly an instant; detected the disturber, then turned and dozed, and dreamt again; if possible, still more disagreeably than before. (…)

p. 25 'I must stop it, nevertheless!' I muttered, knocking my knuckles through the glass, and stretching an arm out to seize the importunate branch: instead of which, my fingers closed on the fingers of a little, ice-cold hand!

The intense horror of nightmare came over me; I tried to draw back my arm, but, the hand clung to it, and a most melancholy voice sobbed,

'Let me in – let me in!' (…)

As it spoke, I discerned, obscurely, a child's face looking through the window – Terror made me cruel; and finding it useless to attempt shaking the creature off, I pulled its wrist on to the broken pane, and rubbed it to and fro till the blood ran down and soaked the bed-clothes: still it wailed, 'Let me in!' and maintained its tenacious gripe, almost maddening me with fear. (…)

p. 26 Thereat began a feeble scratching outside, and the pile of books moved as if thrust forward.

I tried to jump up; but, could not stir a limb; and so yelled aloud, in a frenzy of fright.

To my confusion, I discovered the yell was not ideal [imaginary]. Hasty footsteps approached my chamber door: somebody pushed it open, with a vigorous hand, and a light glimmered through the squares at the top of the bed. I sat shuddering yet, and wiping the perspiration from my forehead: the intruder [Heathcliff] appeared to hesitate and muttered to himself'.

88. *Wuthering Heights*, Vol. I, Chap. III, p. 29.
89. Steiner, op. cit., p. 317.
90. *The International Journal of the Book*, Vol. 4. no. 2, pp. 23-30.
91. See Mireia Aragay's introduction to *Books in Motion* (2005), p. 19.
92. Steiner, op. cit., p. XII.
93. Steiner, op. cit., pp. 437-438:

> Being intermediate and ubiquitous, the great area of 'transformations' and metamorphic repetitions is one in which verbal signs are not necessarily 'transmuted' into non-verbal sign systems. They may, on the contrary, enter into various combinations with such systems.

94. Steiner, op. cit., p. 437.
95. Steiner, op. cit., p. XVI.
96. Steiner, loc. cit.
97. See p. 183 of *Rethinking the Novel/Film Debate* (2003). Literary Cinema in the 'Form/Content Debate' corresponds to the fifth chapter of Kamilla Elliot's monograph.
98. Steiner, op. cit., pp. 312-435.
99. Steiner, op. cit., p. 262.

Part II

100. *The Intimate Screen: Early British Television Drama* (2000), based on his Ph.D. thesis (University of East Anglia).
101. Notably two excellent papers, 'Hitchcock and *The Manxman*: A Victorian Bestseller on the Silent Screen' by Mary Hammond (University of Southampton, 2008) and 'Melodramatic Transformation: George Eliot and the Refashioning of *Mansfield Park*' by Laura Mooneyham White (University of Nebraska-Lincoln), the latter published in *Persuasions: The Jane Austen Journal* (2003).
      See the Bibliography for the full references.
102. *The Intimate Screen*, p. 14.

103. In 1927, Ideal became part of *Gaumont-British Picture Corporation*. In 1941, *Gaumont-British Picture Corporation* was taken over by the *Rank Organisation*. Its extant material has been owned by *Carlton* since 1997.

104. *The History of the British Film (1919-1929)*, pp. 119-120.

105. FIAF: Fédération Internationale des Archives du Film. This 1978 event is widely recognised as having inspired a rethinking of early cinema in general, and early British cinema in particular. See for instance Roger Holman (ed.), *Cinema 1900-1906: An Analytical Study* (Brussels: FIAF, 1982), which disseminated the proceedings of the conference.

    This level of research was soon extended to the later decades of silent cinema.

106. For more details, see the seminal essay written by Sergei Eisenstein in 1944, 'Dickens, Griffith and the Film Today', included in translation in his collection *Film Form* (1949).

    Here is a list of film titles adapted from the works of Dickens (1912-1921):

    Oliver Twist (1912): Thomas Bentley (*Hepworth Manufacturing Company*)
    David Copperfield (1913): Thomas Bentley (*Hepworth Manufacturing Company*)
    The Old Curiosity Shop (1914): Thomas Bentley (*Hepworth Manufacturing Company*)
    Hard Times (1915): Thomas Bentley (*Trans-Atlantic Film Company*)
    Dombey and Son (1917): Maurice Elvey (*Ideal Film Company*) with the script by Eliot Stannard
    Bleak House (1920): Maurice Elvey (*Ideal Film Company*)
    The Old Curiosity Shop (1921): Thomas Bentley (*Welsh, Pearson and Company*)
    The Adventures of Mr Pickwick (1921): Thomas Bentley (*Ideal Film Company*) with the script by Eliot Stannard

107. These films were adapted from the works of Hall Caine (1915-1929):

    The Christian (1915): George Loane Tucker (*London Film Company*)
    The Manxman (1916): George Loane Tucker (*London Film Company*)
    The Prodigal Son (1923): A.E. Coleby (*Stoll Film Company*)
    The Christian (1923): Maurice Tourneur (*Samuel Goldwyn*) US
    The Bondman (1929): Herbert Wilcox (*British and Dominions Film Corporation*)
    The Manxman (1929): Alfred Hitchcock (*British International Pictures*)

108. Extracted from the article entitled 'Ideal's New List: Remarkable Selection of Stories' in *The Bioscope* (10th June, 1920).

    Also, in *The Bioscope* (7th Oct, 1920), still from the *British Library's Newspaper Library*, see the double page addressed to the 'Exhibitors'.

    Finally, refer to 'Ideal Completing the Big Seven. A Batch of Trade Shows' in *The Bioscope* (1st July, 1920).

109. Chap. III, 'Cecil Hepworth, *Alice in Wonderland* and the Development of the Narrative Film' in *Young and Innocent? The Cinema in Britain 1896-1930*, pp. 42-64.

110. *The Yorkshire Post* (April 1920).

111. *Brontë Transformations*, pp. 114-116.

112. *Pelléas et Mélisande* (1892), Maurice Maeterlinck's dramatic masterpiece is a triangular love story, set in the Greek mythology, which recounts the destructive passion of Princess Mélisande, who falls in love with her husband's younger brother, Pelléas.

113. See pp. 46-47 and 452-458 of Miriam Allott's *The Brontës: The Critical Heritage*.

114. See pp. xx-xxi of Pauline Nestor's introduction and pp. 454-455 of Miriam Allott's *The Brontës: The Critical Heritage*.

115. Hoffmann's *Tales* were written in the early decades of the nineteenth century.

116. The première took place at the Opéra Comique, Paris, in 1881.

117. *All Our Yesterdays*, p. 121.

118. His review was published in the *Athenaeum* on 16th June, 1883, then reprinted in *Miscellanies* in 1895. Swinburne had planned and constructed his article as a review of Mary Robinson's full-length biography, *Emily Brontë* (1883).

    Miriam Allott's *The Brontës: The Critical Heritage*, p. 411.

119. The First Mythical Component as defined in the 'Chart of the Mythical Components, Bataillan Themes and Planar/Gothic Figures' or 'Chart of the Dynamic Structures of *Wuthering Heights*'.

120. The Bell Brothers had their first work, *Poems*, released in April 1846.

    Currer Bell's *Jane Eyre* (written by Charlotte) was published in October 1847; Ellis Bell's *Wuthering Heights* (two volumes) and Acton Bell's *Agnes Grey* (one volume) were written by Emily and Anne, respectively, and published in a three-decker version in December 1847.

121. Extracted from her article '*Jane Eyre* and *Wuthering Heights*', which was published in *The Common Reader* in 1925.

122. Virginia Woolf's 1926 essay 'The Cinema' in L. Woolf (ed.) *Collected Essays*, Vol. 2 (1966), pp. 269-270.

123. Steiner, op. cit., p. 438 (Chap. 6 'Topologies of Culture').

124. For a discussion of Yoshida's film, see Philippe Jacquier's interview.

125. *The Bioscope* (5th Aug, 1920): Film Reviews, p. 47.

126. 'Shibden Hall and High Sunderland': Here is the complete transcription of that letter:

Halifax Museums
9th June, 1967

Dear Sir,

Thank you for your letter of the 31st May. I must apologise for not answering earlier.

I do not see any reason why you should not use Shibden Hall in the production of *Wuthering Heights*. But I only control the hall and the immediate garden, the remainder of the Park comes under the Superintendent of the Park's Department and his permission would have to be obtained before any extensive shooting in the park took place.

Of course, as you realise, our greatest problem will be the visitors. It would be impossible at this time of year to close the building as many of our booked parties are arranged months in advance. We also open on Sunday afternoon[s] from 2.30 to 5 pm. These difficulties, I think, could be got over, particularly if you consider filming early in the morning before 11 o'clock.

R.A. Innes,
Museums' Director

Mr. D. Conroy,
Producer BBC2 Classic Serials.
The BBC,
Television Centre,
Wood Lane.
London, W.12

I wonder if you are aware of a legend, or perhaps pious aspiration would be better, in this area. Emily, when she taught at a school some 2 miles from Shibden Hall, did in fact use Shibden Hall as Thrushcross Grange with its close topographical location. And what I think much more remarkable is her utilisation of a now demolished property called 'High Sunderland' for the exterior of Wuthering Heights.

127.   Here are two instances of Joseph's speech in Chap. II, on pp. 9-10:

> 'Whet are ye for?' he shouted. 'T' maister's dahn i' t'fowld [sheepfold]. Goa rahnd by th' end ut' laith [barn], if yah went tuh spake tull him.'

and on p. 17:

> 'Maister, maister, he's staling t'lantern!' shouted the ancient, pursuing my retreat. 'Hey, Gnasher! Hey, dog! Hey, Wolf, hold him, hold him!'

128.   In Chap. XVII, the readers are given a realistic and physical description of the farm's windows as Heathcliff forces his entrance into the house:

The stanchions stood too close to suffer his shoulders to follow, [...] He [then] took a stone, struck down the division between two windows, and sprung in.

129. In Chap. IX, the text reads:

[...] a huge bough fell across the roof, and knocked down a portion of the east chimney-stack, sending a clatter of stones and soot into the kitchen fire.

130. 'WUTHERING HEIGHTS.' ; HINDLEY EARNSHAW RETURNS, MASTER OF 'WUTHERING HEIGHTS.' ; HEATHCLIFF DETERMINES TO LEAVE 'WUTHERING HEIGHTS.' ; EDGAR LINTON AND CATHY ; THE DEATH OF CATHY.
131. Wordsworth's poem, 'I wandered lonely as a cloud' (1807).
132. Eliot Stannard, *Writing Screen Plays* (1920), p. 14.
133. Christine Gledhill, *Reframing British Cinema (1918-1928)* (2003), pp. 119 and 122.
134. Gledhill, op. cit., pp. 51-52 and pp. 93-95, respectively.
135. Gledhill, op. cit., p. 93.
136. Charles M. Berg's 'The Human Voice and the Silent Cinema' in the *Journal of Popular Film*, pp. 174-175.
137. Tom Gunning's Chap. 8, in *The Silent Cinema Reader* (2004), p. 150.
138. Noël Burch's *To the Distant Observer: Form and Meaning in Japanese Cinema*, p. 79.
139. Christine Gledhill, 'Coda: Hitchcock, *The Manxman* and the Poetics of British Cinema', p. 122 (last paragraph).
140. Jon Burrows's essay, 'It would be a Mistake to strive for Subtlety of Effect: Richard III and Populist, Pantomime Shakespeare in the 1910s', in *Young and Innocent?*, p. 83.
141. Christine Gledhill's *Melodrama and Realism in 1920s British Cinema*.
142. *The Seven Lively Arts* (Mineola, New York: Dover Publications, 2001; first published 1924).
143. *The Art of the Moving Picture* (New York: The Macmillan Co., 1922; first published 1915).
144. For a brilliant study of Montagu's intellectual credos and a thorough investigation into the achievements of the *Film Society*, see the complementary chapters written by Gerry Turvey, 'Towards a Critical Practice: Ivor Montagu and British Film Culture in the 1920s' (pp. 306–320), and by Jamie Sexton, 'The Film Society and the Creation of an Alternative Film Culture in Britain in the 1920s' (pp. 291-305), in *Young and Innocent?*
145. Kiju Yoshida's *Ozu's Anti-Cinema* (2003), pp. 148-149.
146. See Ian Macdonald's article, 'The Silent Screenwriter: The Re-discovered Scripts of Eliot Stannard' (2008).
147. Jon Burrows' essay, 'It would be a Mistake to strive for Subtlety of Effect: Richard III and Populist, Pantomime Shakespeare in the 1910s', p. 89.

148. Gerry Turvey's 'Enter the Intellectuals: Eliot Stannard, Harold Weston and the Discourse on Cinema and Art' in Alan Burton and Laraine Porter's *Scene-stealing. Sources for British cinema before 1930* (2003), pp. 85-93.

149. Ian Macdonald's article, 'The Silent Screenwriter: The Re-discovered Scripts of Eliot Stannard' (2008).

150. François Truffaut's *Hitchcock* (1967).

151. See the *Supplement to The Bioscope* of 1ˢᵗ July, 1920, and the piece entitled 'British Studios' which is dedicated to Ideal Films, Ltd.

152. This biography featured amongst the documents that the collector (and scholar) Fred Lake made available to Ian Macdonald for his research on Eliot Stannard.

153. From the aforementioned biography.

154. Charles Barr's article, '*Blackmail:* Silent and Sound' in *Sight and Sound* (Spring 1983), pp. 189–193.

155. Macdonald's 'The Silent Screenwriter'.

156. From the aforementioned biography.

157. See the *Monthly Film Bulletin* review (Vol. 19, no. 216, January 1952) available on the BFI screenonline website.

158. Rachael Low's *The History of the British Film 1919-1929*, pp. 87 and 118-119, respectively.

159. Low, op. cit., p. 119.

160. Low, op. cit., p. 274.

161. Luke McKernan's chapter entitled 'Shooting Stars' in *The Cinema of Britain and Ireland* (2005), pp. 11-18.

162. McKernan, op. cit., pp. 13-15.

163. Charles Barr's 'Writing Screen Plays', p. 229.

164. *The Bioscope* review, pp. 65-66.

165. Stannard's films at Gainsborough with Hitchcock:

   The Pleasure Garden (1927)
   The Mountain Eagle (1927)
   The Lodger (1927)
   Downhill (1927)
   Easy Virtue (1927)

   Stannard's films at BIP with Hitchcock:

   The Ring (1927)
   Champagne (1928)
   The Farmer's Wife (1928)
   The Manxman (1929)

166. See 'The Man Who wasn't There', pp. 18-19:

Stannard was one of the very first practitioners of the profession to think seriously and write publicly about exactly what it takes to be a writer for the screen. And his articles about the craft remain, to my mind, more analytical and useful than the shelves of 'how to' volumes that have proliferated in recent years. [...]

Stannard's first principles: the scenario writer must be 'thoroughly experienced in each technical branch of Kinematography; possessed of dramatic training and a sense of Theatre; conversant with the laws of literary construction; a student of psychology and character; and alive to the atmospheric value of costume, furniture, architecture and scenery.' How many of today's script editors (whose companies have stumped up for them to attend a weekend seminar by guru Robert McKee) could admit themselves adept at even one of these five essential requisites? How many writers could? [...]

There's no Blue Plaque, no biography. Most of the 116 films for which he has a credit are missing, believed dead. Marginal, ephemeral, unappreciated, obliterated, now is the hour for Eliot Stannard to be honoured as the patron saint of British screenwriters.

167. John Boorman and Walter Donohoe's *Projections 2* (1993), p. 123.
168. 'Symbolism', *Kinematograph and Lantern Weekly*, 23rd May, 1918, p. 76
    'The British Film Actor', *Kinematograph and Lantern Weekly*, 30th May, 1918, p. 79
    'The Unseen', *Kinematograph and Lantern Weekly*, 6th June, 1918, p. 97
    'The Life of a Film', *Kinematograph and Lantern Weekly*, 13th June, 1918, p. 81
    'What of the Future', *Kinematograph and Lantern Weekly*, 20th June, 1918, p. 8

169. Stannard's *Writing Screen Plays* (1920), p. 14.
170. Steiner, op. cit., p. 438.
171. A 'Deemster' is a high-ranking judge (of whom there are only two) on the Isle of Man.
172. *Wuthering Heights*, Vol. I, Chap. X, p. 97.
173. *Wuthering Heights*, Vol. II, Chap. III, p. 184 and Vol. II, Chap. IV, p. 189.
174. 'Coda: Hitchcock, *The Manxman* and the Poetics of British Cinema', p. 119.
175. See Jonathan Bignell's 'The Spaces of The Wednesday Play (BBC TV 1964-1970): Production, Technology and Style', p. 373.
176. *The Manxman*'s illustrations by Fred Pegram appeared in the *Queen* serialisation of 1894 (between January and July). See Mary Hammond's essay that is now part of Barton Palmer and David Boyd (eds) *Hitchcock at the Source* (New York: SUNY, 2011).
177. Both quotes are taken from Mark Duguid's *Screenonline* review of *The Manxman*.
178. Thomas Leitch's 'The Adapter as Auteur: Hitchcock, Kubrick, Disney' in Aragay's *Books in Motion*, pp. 107-121.

Part III

179. 'Eliot Stannard on the Cameraman and the Scenario Writer', account of the lecture to the *Kine-Cameramen's Society*, *Kinematograph Weekly* Supplement, 17th Feb, 1921, p. xx:

> [...] a kind of trinity composed of the producer, scenario-writer, and cameraman [...] the cameraman would become a creative artist working in collaboration with his two *confrères* [...]

180. A. Scott Berg's *Goldwyn: A Biography* (1989), pp. 319-320.

181. Jacques Doniol-Valcroze and André Bazin's 'Entretien avec Luis Buñuel' (pp. 10-11) in *Cahiers du Cinéma*, no. 36 (June 1954) cited and translated in Francisco Aranda's *Luis Buñuel: A Critical Biography* (1969), p. 162.

182. Pierre Unik had first collaborated with Buñuel on the commentary of *Las Hurdes* (*Land Without Bread*, 1932), a 27-minute long documentary still shown today all over the world.

183. Luis Buñuel's *My Last Breath* (1982), p. 111.

184. Since 1968, the 'Prix Georges Sadoul' has been awarded to the best feature films made by new French or foreign film-makers. It has to be their first or second film only.

185. Francisco Aranda's *Luis Buñuel: A Critical Biography*, pp. 97-98.

186. Luis Buñuel's *My Last Breath* (1982), p. 205.

187. Luis Buñuel, José de la Colina, Tomás Pérez Turrent's *Conversations avec Luis Buñuel: Il est dangereux de se pencher au-dedans* (*Luis Buñuel: Prohibido asomarse al interior*, 1986), translated by Marie Delporte and Charles Tesson (1993), p. 116.

188. See the scene of Cathy's delirium in Emily Brontë's novel, Vol. I, Chap. XII, p. 125:

> You may fancy a glimpse of the *abyss* where I grovelled! [Emphasis added]

189. In 1939, Buñuel was hired by Iris Barry and worked in the Film Department of the *Museum of Modern Art* at the selection, re-editing and dubbing of some English (or North American) documentaries destined to the South American market.
    See Francisco Aranda's *Luis Buñuel* (1969), p. 122.

190. Aged only sixteen, she started her film career in Vittorio de Sica's *Maddalena, zero in condotta* (*Maddalena: Zero for Conduct*, 1940).

191. *My Last Breath* (1982), p. 205.

192. Doniol-Valcroze and Bazin's 'Entretien avec Luis Buñuel' (p. 11) in *Cahiers du Cinéma*, no. 36 (1954):

> Tous les autres [films] je les ai faits en vingt-cinq jours de tournage.

193. José Francisco Aranda's *Luis Buñuel* (1969), p. 98:

> When everything was prepared, however, the project aborted because the studios were destroyed by fire.

194. See José de la Colina and Tomás Pérez Turrent's *Conversations avec Luis Buñuel* – which were compiled after Luis Buñuel's death in Mexico in 1983.

195. *Vaghe stelle dell'orsa* – released as *Of a Thousand Delights* in the United Kingdom.

196. *Wuthering Heights*, Vol. I, Chap. X, p. 93.

197. Steiner, op. cit., Chap. 5 'The Hermeneutic Motion', p. 314.

198. 3rd Mar, 1845.

199. The character is, in actual fact, called 'Catalina'.

200. See the scene of Cathy's delirium in Emily Brontë's novel, Vol. I, Chap. XII, p. 126:

> I'll not lie there by myself, they may bury me twelve feet deep, and throw the church down over me; but I won't rest till you are with me ... I never will!

201. Anthony Fragola's 'Buñuel's Re-vision of *Wuthering Heights*' in *Literature/Film Quarterly* (1994), p. 52.

202. *Wuthering Heights*, Vol. II, Chap. XX, p. 335:

> Mr Heathcliff was there – laid on his back. His eyes met mine so keen, and fierce, I started; and then, he seemed to smile.
>
> They [His eyes] would not shut [...] and his parted lips, and sharp, white teeth sneered too! [...]
>
> 'Th' divil's harried off his soul,' he cried, 'and he muh hev his carcass intuh t'bargin, for ow't Aw care! Ech! what a wicked un he looks girnning at death!' and the old sinner grinned in mockery.

203. Michael Popkin's 'Wuthering Heights and Its Spirit' in *Literature/Film Quarterly* (1987), p. 121.

204. This blasphemous passage is excerpted from the *Book of Wisdom*, Chap. II, verses 1–9. According to Buñuel, 'the author had to put these words into the mouths of unbelievers in order to get them printed'.
    Luis Buñuel's *My Last Breath* (1982), pp. 205-206.

205. Julie Jones' 'Fatal Attraction: Buñuel's Romance with *Wuthering Heights*' in *Anales de la Literatura Española Contemporanea* (1997), p. 158.

206. Anthony Fragola's 'Buñuel's Re-vision of *Wuthering Heights*' (1994), pp. 53-54.

207. This discussion of the spider insert with Buñuel is part of the interview involving José de la Colina, Tomás Pérez Turrent and Luis Buñuel in *Conversations avec Luis Buñuel* (p. 118). It is subsequent to Buñuel's interview with Doniol-Valcroze and Bazin in 1954.

208. Michael Popkin's 'Wuthering Heights and Its Spirit' (1987), pp. 119-120.

209. Carlos Córdova's *Agustín Jiménez y la vanguardia fotográfica* (2005).

210. See Olivier Debroise and Stella de Sa Rego's *Mexican Suite: A History of Photography in Mexico* (2001), pp. 141–142:

> At the hacienda of Tetlapayac where Eisenstein was filming *¡Qué viva México!*, Agustín Jiménez took a series of famous portraits of the Soviet film-maker playing with a 'Day of the Dead' sugar skull. It is believed that Jiménez was strongly influenced by the Russian photographers and by Eisenstein's ideas on composition.

211. Francisco Aranda's *Luis Buñuel* (1969), p. 163:

> He [Buñuel] side-steps the actors, using them as he uses the skeletal trees in the dark and jagged landscape, as objects in a composition in which the camera and the editing are the true protagonists. By faithfully realizing his project of twenty years earlier […] he achieved an anachronistic quality, a little like some old film by Dmitri Kirsanov, which only adds to the fascination of the work.

212. Kamilla Elliott's 'Literary Cinema in the Form/Content Debate' in *Rethinking the Novel/Film Debate* (2003), p. 152.

213. Gaston Bachelard's *La Terre et les rêveries de la volonté* (1947) and *La Terre et les rêveries du repos* (1948).
   See also Mircea Eliade, *Myths, Dreams and Mysteries* (1957), Part IV, p. 79:

> In the paradisiac age, the gods came down to Earth and mingled with men; and men, for their part, could go up to Heaven by climbing the mountain, the tree, creeper or ladder, or might even be taken up by birds.

214. *Wuthering Heights*, Vol. I, Chap. XI, p. 108.

215. Anthony Fragola's 'Buñuel's Re-vision of *Wuthering Heights*' (1994), p. 54.

216. See Rivette's interview.

217. *Viridiana* (1961) – Benito Pérez Galdós' *Halma* (1895); the script co-written with Jean-Claude Carrière (1965) of *Le Moine* (1972) – Matthew Gregory Lewis' *The Monk* (1796); *Belle de Jour* (1966) – Joseph Kessel's *Belle de Jour* (1928); *Tristana* (1970) – Galdós' *Tristana* (1892); and *That Obscure Object of Desire* (1977) – Pierre Louÿs' *La femme et le pantin* (1898).

218. The cinematographers Gabriel Figueroa, José Ortiz Ramos and Raúl Martínez Solares also crystallise around Buñuel's Mexican years (from 1946 to 1965).

219. Berg, loc. cit.

220. Francisco Aranda's *Luis Buñuel: A Critical Biography* (1969), p. 117.

221. Jan Herman's *A Talent for Trouble: William Wyler* (1995), p. 289.
222. Michel Ciment and Christian Viviani's 'Dossier William Wyler' in *Positif* (June 2001), pp. 68-102.
223. Berg, op. cit., p. 324.
224. Berg, op. cit., p. 321.
225. Gregg Toland suffered a premature death in 1948. He was only aged 44.
226. Herman, op. cit., pp. 143-144.
227. Pamela Mills' 'Wyler's Version of Brontë's Storms in *Wuthering Heights*' in *Literature/Film Quarterly* (1996), p. 321, and Berg, op. cit., p. 418.
228. Berg, op. cit., p. 324.
229. *Behind the Camera: The Cinematographer's Art* (1971), p. 34.
230. Brian McFarlane's *Novel to Film* (1996), pp. 13-14.
231. George Bluestone's '*Wuthering Heights*' (1957), p. 105.
232. Mills, op. cit., p. 415.
233. Berg, op. cit., p. 326.
234. Berg, op. cit., p. 328.
235. At the 1939 Academy Awards, Geraldine Fitzgerald was nominated as Best Actress in a Supporting Role.
236. John Gassner and Dudley Nichols' *Twenty Best Film Plays* (1943), p. 322.
237. John Harrington's 'Wyler as Auteur' in *The English Novel and the Movies* (1981), p. 80.
238. Gassner and Nichols, op. cit., p. 324.
239. Berg, op. cit., p. 326.
240. Although it did not receive the Best Picture Academy Award, the prestigious New York Film Critics Circle thought it was indeed the best picture of the year.
241. See Amy Martin's article 'A Battle on Two Fronts: Wuthering Heights and Adapting the Adaptation' in *Film Remakes, Adaptations and Fan Productions: Remake/Remodel* (2012).
242. Berg, op. cit., p. 328.
243. 'Reviews for Showmen' in *Kinematograph Weekly*, 27th Apr, 1939, p. 28.
244. 1953 BBC Audience Research Report.
245. 'Quebec Tightening Up on Films' in *Variety*, 12th Apr, 1939, p. 7.
246. C.H. Lemon's 'Film Review: *Abismos de pasión* – A Film based on *Wuthering Heights*' in the *Brontë Society Transactions* (1984), Vol. 18, no. 4, p. 310.
247. Seton Margrave wrote *Meet the Film Stars* (London, 1934-1935) and also *Successful Film Writing* (1936).
248. *The Daily Mail*, Friday 28th Apr, 1939, p. 18.
249. Back in 1939, the 1920 *Wuthering Heights* may still have been in the archives of the *Gaumont-British Picture Corporation*. However, because so many silent films were being routinely discarded as junk when synchronised sound came in, it may as well have already gone missing by then.
250. 1978 BBC Audience Research Report (First Episode).

251. George Bluestone's *'Wuthering Heights'* in *Novels into Film* (1957), p. 99.
252. Gassner and Nichols, op. cit., p. 309.
253. Steiner, op. cit., p. 317.
254. Gassner and Nichols, op. cit., p. 300.
255. *Wuthering Heights*, Vol. I, Chap. VII, p. 58.
256. Mills, op. cit., p. 418.
257. Steiner, op. cit., p. XVI.
258. Steiner, op. cit., p. 318.
259. Mills, op. cit., p. 419.
260. In Episode 15, Season 2.
261. The 1948 Teleplay: Memorandum Addressed to George More O'Ferrall.

From: Mr Robert McDermot
Subject: WUTHERING HEIGHTS
To: Mr. George More O'Ferrall

8th Mar, 1948

I should like to put on paper my compliments on your production of this undoubtedly bad play, and to let you know that, in common with the critics, I felt that this and the acting were responsible for any success the play may achieve with viewers. At the same time I think you did cut too much of it. As the play itself is of such a corney [corny] nature I think viewers, if they liked it at all, would have been quite prepared to accept a good deal more of the admittedly hamy [hammy] dialogue, and that by cutting so ruthlessly you made the story itself a little difficult to follow for those who were not already family [familiar] with it. The camera work and the smooth running of the production were altogether admirable, and the point I have just made is my only criticism.

Robert McDermot

262. Nigel Kneale's 'An Electrifying Masterpiece', *Radio Times*, 4th Dec, 1953.
263. The full original statement that Mr Elwyn Jones, BBC Assistant Head of Drama, had to make at New Scotland Yard, on 13th June, 1962, is available at the BBC WA.
264. *This is a letter written by Nigel Kneale to Robin Wade in response to the letter that Wade had sent to him on 18th Jan, 1962.*

17, Holland Park,
W.11

31st Jan, 1962

Dear Robin,

Thanks for the WUTHERING HEIGHTS script, returned herewith. I remember this one well... I wrote it in 8 days flat, the whole production being thought of, cast and performed suddenly upon the unexpected appearance of Mr Richard Todd and his desire to play Heathcliff. A sort of midget Heathcliff. Interesting. Particularly a scene he played currycombing the hind legs of a prop horse that had no forepart and rang hollow when slapped. (Can that be revived too, I wonder?)

I've read it through. I don't think it's at all bad for one written in 8 days. The shape's all right and the action works. It's only the first half of the book, of course, but if you put the rest in you have to do it as a serial. Dialogue... well, it fits those Freudian abstractions that Miss Brontë gave names [to] and pretended were characters. (Claire Bloom as the wild sprite of the Yorkshire Moors? Crumbs.)

I've marked a number of literals which actors have a weird tendency to pronounce unperturbed, and one or two dud lines. (Not to say there aren't others, but a script is like knitting: once you start unravelling...!)

Credit? I think not, thanks. It was a long time ago, and I'd probably do it differently today, even in 8 days. Perhaps it's best as an anonymous BBC version?

The moving hasn't hit us yet. We've bought, and now have builders in, God help us. Theoretically we move at the end of March, but ...What are those LCC hostels like, I wonder.

Best wishes from us all,
Yours, Tom (Nigel Kneale)

Robin Wade, Esq. ,
Senior Assistant, Script Dept.
BBC Television.

265. *Wuthering Heights*, Vol. II, Chap. XX, p. 337.
266. A Rare Memorandum from Donald Baverstock:

The play WUTHERING HEIGHTS on Friday, 11th May, received an astonishingly large audience. We discussed why this could be so at Programme Planning Committee.

Various explanations were made. Firstly, that it has a strong title and a strong star; that it is a popular classic, 'the ignorance of which everyone wants to get off their conscience' and that 'the opposition to it [on ITV] was not particularly strong'.

I myself was not convinced that any of these reasons sufficiently explained the remarkable size of this audience. I should be grateful, therefore, if you would give us your explanation of it. It may be that there are very important lessons for us to bear in mind in connection with future planning.

267. Peter Holdsworth's article 'Another *Wuthering Heights Film*' was published in the *Telegraph and Argus* (p. 8) on Thursday 2nd July, 1964.

268. This letter can be found in the Lindsay Anderson Collection (5/4/1) Unrealised film projects – Miscellaneous projects (1950s to 1960s) where it is referenced as number 20.

The Lindsay Anderson Collection is curated by the 'Archives and Special Collections' of the University Library of Stirling.

269. These notes are referenced as 5/4/1/20.

270. *Wuthering Heights*, Vol. II, Chap. XVIII, p. 305.

271. *Wuthering Heights*, Vol. II, Chap. XIX, p. 324.

272. Extract from Villiers de l'Isle-Adam's *Véra* (1874) in his *Contes cruels*:

> In the end, by reason of the deep and all-compelling will of d'Athol, who thus from the strength of his love wrought the very life and presence of his wife into the lonely mansion, this mode of life acquired a gloomy and persuasive magic. [...]
> The glimpse of a black velvet robe at the bend of a pathway; the call of a laughing voice in the drawing-room; a bell rung when he awoke in the morning, just at it used to be – all this had become familiar to him: the dead woman, one might have thought, was playing with the invisible, as a child might.

273. *Wuthering Heights*, Vol. II, Chap. XX, p. 331.

274. Jonathan Bignell's 'The Spaces of the Wednesday Play (BBC TV 1964–1970): Production, Technology and Style' (2014), p. 374.

275. See the complaint letter that David Conroy addressed to the Head of Scenic Servicing on 30th Aug, 1967:

> Some pieces of scenery never arrived at the studio and make-shifts had to be created on the floor. Bad handling between Alexandra Palace and Riverside studio of the stock sets caused pin hinges to be broken and door hinges to be lost.

as well as Peter Sasdy's thank you note to Peter Kindred dated 22nd Sept, 1967:

> Many thanks for all the hard work you put into the four productions of "Wuthering Heights". It is a great pity that we could not do the first two at the Centre as well as the last – but that's how it goes!

276. Bignell, op. cit., p. 379.

277. From the original memo written by Peter Sasdy, the serial director, 26th June, 1967. See also Shaun Sutton's The Largest Theatre in the World: Thirty Years of Television Drama (1982).

278. From the original note written by Ian Strachan, one of Peter Sasdy's Production Assistants, on 9th Aug, 1967.

279. Sasdy's letter to Conroy (19th June, 1967).

280. Conroy's letter to Leonard (10th Aug, 1967).

281. Letter from Carolyn Bill to Hugh Leonard (10th July, 1967).

282. 30th May, 1967.

283. BBCWA T5/704/2.

284. Letter dated 31st May.

285. Bignell, op. cit., p. 375.

286. Bignell, op. cit., p. 375.

287. From an unsigned memo addressed to Christopher Hayden, Wyvern Productions Ltd.

288. Bignell, op. cit., pp. 376-377.

289. See 'Definitions, Early History: The Classic Drama Serial' (pp. 19–23 and p. 26) in Robert Giddings and Keith Selby's *The Classic Serial on Television and Radio* (2001).

290. Chap. XIV, Samuel T. Coleridge's *Biographia Literaria* (1817).

291. Bignell, op. cit., p. 383.

292. *Wuthering Heights*, Vol. II, Chap. II, pp. 168-169.

293. Andrea, in her interview with Benjamin Secher, declared:

> You haven't seen half of the mud […]. In fact when I watched the early footage I felt really annoyed. It doesn't truly show how hard that environment was. It was unbelievably tough for everybody.

Refer to Secher's article in *The Telegraph*, 'Dark Depths of Andrea Arnold's *Wuthering Heights*' (Saturday 5th Nov, 2011).

294. Peter (Charles) Hammond is still well remembered for his prominent part in *The Adventurers* (1951), a British adventure film made for the big screen, as well as for his swashbuckling role in *The Buccaneers* (1956–1957) and *William Tell* (1958–1959), both television series.

295. Peter Sasdy, who was twelve years Peter Hammond's junior, started his directorial career early with the high-profile teleplays *The Caves of Steel* (1964) and *The Stone Tape* (1972), and went on capitalising on his directorial experience with Hammer Films in the early 1970s with such popular series as the *Hammer House of Horror* (1980) and the *Hammer House of Mystery and Suspense* (1984), for instance.

296. *Wuthering Heights*, Vol. I, Chap. XI, pp. 116-119.

297. *Wuthering Heights*, Vol. II, Chap. II, pp. 169-170.

298. Dick Fiddy's 'Peter Hammond: Actor Who Became a Prolific TV Director' in *The Guardian* (Sunday 1st Jan, 2012).
299. *Wuthering Heights*, Vol. I, Chap. XII, pp. 121-123.
300. *Wuthering Heights*, Vol. II, Chap. XV, pp. 288-291.
301. Christine Gledhill's sub-chapter 'Pictorialism and Modernity' in *Reframing British Cinema (1918–1928): Between Restraint and Passion*, p. 52.
302. Apologies for using the 'f' word of Adaptation Studies here but I could not find a more appropriate epithet ...
303. *Wuthering Heights*, Vol. II, Chap. XVII, p. 299.
304. *Wuthering Heights*, Vol. II, Chap. XVIII, pp. 312-316.
305. *Wuthering Heights*, Vol. I, Chap. XIII, p. 135.
306. See Bob Fuest's interview.
307. Patrick Tilley's response to Mrs Hanna Taussig in *The Times* (18th Apr, 1970).
308. *Wuthering Heights*, Vol. I, Chap. VIII, p. 64.
309. *Wuthering Heights*, Vol. I, Chap. IV, p. 36.
310. See Patrick Tilley's interview.
311. See Bob Fuest's interview.
312. Steiner, op. cit., Chap. 5 'The Hermeneutic Motion', pp. 314-315.
313. Rick Schmidt's '*Wuthering Heights*' in *Variety* (16th Dec, 1970).
314. Brontë Society's *Transactions* (1975), Vol. 16, p. 59.
315. Schmidt, op. cit.
316. 'Wuthering Depths' in the *New York Magazine* (22nd Feb, 1971), p. 68.
317. Patrick Tilley's interview.
318. Patrick Tilley's '*Wuthering Heights*' in *The Times* (18th Apr, 1970).
319. Patrick Tilley's interview.
320. Wimbledon School of Art.
321. For the role of Field Marshal Sir John French, Laurence Olivier received the BAFTA Film Award for Best Supporting Actor in 1970.
322. For his work on the film, Gerry Turpin received the *Golden Globe* and the BAFTA Film Award for Best Cinematography in 1969 and 1970, respectively.

    Moreover, Donald M. Ashton received the BAFTA Film Award for Best Art Direction, Anthony Mendleson for Best Costume Design and Don Challis and Simon Kaye for Best Sound Track (1970).
323. Several of Len Deighton's novels (such as *The IPCRESS File* or *Funeral in Berlin*) were adapted to the screen. The films based on these novels featured an anti-James Bond hero named Harry Palmer and portrayed by Michael Caine.
324. Here, Patrick makes a slight factual mistake. To retain a certain degree of control on the ending, Bob eventually called back Anna and Timothy and shot the best he could of the 'tear jerker' finale that had been imposed by the producer(s).
325. Marcel Pagnol (1895-1974) was a French novelist and film-maker. In 1946, he became the first film-maker ever to be elected to the Académie Française.

The novels that Patrick is referring to (*Jean de Florette* and *Manon des Sources* in *L'Eau des collines*, 1964) were first adapted by Pagnol himself (1952) then by Claude Berri (1986).

326. Joseph declares sarcastically that 'he [Edgar] is waiting to see the colour of its eyes'.

327. In 1969, he was just finishing off the seven-episode series of *The Avengers* (1968-1969).

328. One of his more recent contributions was the co-writing of the script of Richard Attenborough's *Chaplin* (1992).

329. 1970.

330. 1971. It was released as *Mr Forbush and the Penguins* and directed by Arne Sucksdorff and Alfred Viola.

331. 2002.

332. He composed the scores of most of Alfred Hitchcock's pictures but also, at his debut in 1941, the score of *Citizen Kane*, more recently of *Taxi Driver* (1976), and *Kill Bill: Part 1* (2003) posthumously.

333. Judith Crist in the *New York Magazine* (22nd Feb, 1971).

334. It was *Male of the Species* (1969) with Michael Caine and Sean Connery appearing as guests. The narrator was the inescapable Laurence Olivier.

335. According to the IMDb database, Peter Hammond directed nineteen episodes of *The Avengers* (1961–1964) and Bob Fuest worked as a production designer with Peter on eight of those episodes (January to August 1961). Around the same time, Bob continued as a production designer on another two episodes (December 1961 and 1962) directed by Don Leaver and Richmond Harding, respectively.

336. *Spring and Port Wine* (1970).

337. *Wuthering Heights*, Vol. I, Chap. 1, p. 5.

338. Lucasta Miller and Cynthia Rose's 'How Cathy Came home' in *New Stateman & Society*, 16th Oct, 1992, p. 34.

339. David Skynner's interview.

340. David Skynner's interview.

341. Sara Martín's 'What does Heathcliff Look Like? in Mireia Aragay's *Books in Motion* (2005), p. 62.

342. Martín, op. cit., pp. 63-64.

343. Peter Conrad, 'Jane Eyre and Wuthering Heights: Do We Need New Film Versions?' in *The Observer* (Sunday 21st Aug, 2011).

344. 'The All-Important Archeo-Cinematic Scene' in Kiju Yoshida *Ozu's Anti-Cinema* (2003), p. 30.

345. Yoshida, op. cit., p. 35.

346. Martín, op. cit., p. 56, Footnote 2.

347. According to the *BARB* Report, there were 7.67 million viewers for the 1998 *Wuthering Heights* broadcast on ITV1 on Sunday, 5th Apr. It had the quality of a cinema movie, which would not have surprised Nigel Kneale.

Nigel Kneale's 'Not Quite So Intimate' in *Sight and Sound*, Vol. 28 (1959), p. 86.

348. See Lockwood's narrative in *Wuthering Heights*, Vol. I, Chap. I, p. 5:

> He [Mr Heathcliff] is a dark-skinned gypsy in aspect, in dress and manners a gentleman [...]

And Nellie's cue in *Wuthering Heights*, Vol. I, Chap. VII, p. 58:

> You're fit for a prince in disguise. Who knows, but your father was Emperor of China, and your mother an Indian queen [...]?

349. See David Skynner's interview.
350. Martín, op. cit., p. 64.
351. Here, this is the face of Sinéad O'Connor.
352. Q.D. Leavis regarded the embedded narratives and numerous flashbacks of Emily Brontë's novel (which characterise the Gothic Figure of *Mise en Abîme*) as 'awkward' and very 'confusing'. ('A Fresh Approach to *Wuthering Heights*', 1969).

    The novel's embedded structure is reflected in the film by the presence of the author, Emily Brontë, which draws the attention of the audience to the fictional nature of the setting (and characters) and allows the story to be told in a flashback. In this film version, Nellie and Lockwood are characters only: they do not have a narratorial role, the fictional Emily Brontë has.

353. Mary Selway (1936–2004) was a most-esteemed British casting director. She cast more than a hundred feature films and worked regularly with the likes of Ridley Scott, Steven Spielberg, John Boorman, Sidney Pollack, Robert Altman, Peter Weir and Roman Polanski. On the occasion of *Emily Brontë's Wuthering Heights*, Paramount appointed her as a producer.

354. See p. 4 of the first of the two Film Education Study Guides on *Emily Brontë's Wuthering Heights* that were commissioned by the now extinct Paramount British Pictures.

355. See pp. 5-7 of the first of the two Film Education Study Guides on *Emily Brontë's Wuthering Heights*:

> Before starting to write the script, Devlin got to know the novel very well through a number of readings. [...] She was searching for potential 'filmic' images which the plot could use and discovered *references to branches of trees and hands*. [Emphasis added]

356. Anne Devlin is a Belfast-born writer and scholar. Back in 1990, it was her first experience of film writing. But in 1988, she had adapted D.H. Lawrence's *The Rainbow* for the *BBC* and, by the time she was chosen for *Wuthering Heights*,

already had a trail of literary awards behind her – the Hennessy Literary Award (1982), the Samuel Beckett Award (1984) and the Susan Smith Blackburn Prize (1986). Subsequently, she was awarded the Lloyds Playwright of the Year for her play *After Easter* (1994).

> Consistent with all the drafts is the fact that Emily starts off the story. [...] This follows from the brief given to Devlin by Selway and Maisel. The first draft was achieved and then a second. But the third draft was the most important as it was one written in conjunction with the director. [...] The script itself was in the process of [being written] writing from August 1990 till June 1991 and from December 1990 onwards, more than one voice had a say in what would eventually appear on the screen.

357. See p. 5 of the first of the two Film Education Study Guides on *Emily Brontë's Wuthering Heights*.

358. *The Falklands War: The Untold Story* (1987); *Afghantsi* (1988); *Murder in Ostankino Precinct* (1989); and *One Day in the Life of Television* (1989).

359. It sued American International Pictures over its use of the title of the Samuel Goldwyn movie – funny to think that the title of the Goldwyn movie can take precedence over the title of the novel!

360. The silent way in which Cathy is being portrayed in that scene probably comes from Anne Devlin's script:

> Emily's story is told in and through silence. These silences – the wind opening a door – make us enter into the story. The narrative links are these silences, emphasising the quietness of her life. (See p. 6 of the first of the two Film Education Study Guides on *Emily Brontë's Wuthering Heights*.)

361. Her voice seems to materialise the transition between the first part (first generation, Chap. I to XVI) and the second part (second generation, Chap. XVII to XXXIV) of Emily Brontë's novel.

362. *The Detective* (1985) and *The Life and Loves of a She-Devil* (1986), for instance.

363. *Prime Suspect* (1991-1993 and 1995) and *Cracker* (1993-1995) amongst others.

364. The British Society of Cinematographers.

365. In 2001, Neil McKay was nominated (together with the producer Mike Dormer and the director David Richards) for the BAFTA Television Award of the Best Drama Serial for *This Is Personal: The Hunt for the Yorkshire Ripper* (2000).
    In 2007, he won the Writers' Guild of Great Britain Award for Best Original Drama and the BAFTA Television Award of the Best Drama Serial for *See No Evil: The Moors Murders* (2006).

366. Louise Berridge, started her television career as a script editor for the drama series *Medics* and *Eastenders*. She came back to *Eastenders* as an executive producer between 2002 and 2004.

367. More recently, he played John Birt in the feature *Nixon/Frost* (2008) and Arthur Clennam in the television serial *Little Dorrit* (2008).

368. In 1995, he had interpreted the high-profile Darcy in the BBC's successful serial *Pride and Prejudice*.

369. Naveen Andrews first became well known with the television series, *The Buddha of Suburbia* (1993). He also played Kip, the mine-clearing expert with whom Hana (alias Juliette Binoche) falls in love in *The English Patient* (1996). A little more recently, he starred as Mr Balraj (Mr Bingley) in *Bride and Prejudice* (2004).

370. See Lockwood's narrative in *Wuthering Heights*, Vol. I, Chap. I, p. 5:

> He [Mr Heathcliff] is a dark-skinned gypsy in aspect, in dress and manners a gentleman […].

And Nellie's cue in Wuthering Heights, Vol. I, Chap. VII, p. 58:

> You're fit for a prince in disguise. Who knows, but your father was Emperor of China, and your mother an Indian queen […].

371. 1771-1803.

372. Timothy Dalton was twenty-four years old in 1969, and *Sat'day While Sunday* dates from 1967 (Thames Television).

373. I interviewed Patrick Tilley in January 2003 at his farmhouse in Penyrallt, Wales.

374. I interviewed Robert Fuest at his cottage in Winchester in May 2003.

375. For Lockwood's nightmare scenes, Neil McKay must have used Anne Devlin's idea since the 1998 film narrative is also framed by the two complementary nightmare sequences, one at the beginning and the other one at the end.
    Such a structure helps bring closure to a story that is told almost entirely in flashback and gives a reassuring impression of resolution to the audience.

376. It is probably Ponden House.

377. Robert Shaw played the boat owner, Quint, in Steven Spielberg's *Jaws* (1975).
    He was also a prolific writer in the 1960s. His novel, *The Sun Doctor* (1961), was awarded the Hawthornden Prize in 1962 and was followed by a controversial trilogy, *The Flag* (1965), *The Man in the Glass Booth* (1967) and *A Card from Morocco* (1969).

378. On Sunday 5th Apr, 1998, at 8.00 pm.

379. The official figures of the BARB report read 9.65 at the start and 7.79 million at the end of the programme, which are good figures.

380. 'From Wuthering Heights to the Depths of Despair' in *The Stage* (16th Apr, 1998).

381. 30th to 31st Aug, 2009.

382. Kate McMahon, 'The Broadcast Interview: Laura Mackie, ITV Drama', *Broadcast Now* (25th Feb, 2009).

383. Stuart Jeffries' 'Sex and Rebellion: *Desperate Romantics* Writer Peter Bowker on His New BBC Drama' in *The Guardian* (Tuesday 21st July, 2009).

384. Bataille, op. cit., pp. 5-6.

385. This, however, was not an issue for the North American viewers of PBS Masterpiece Classic, who first saw the Bowker–Giedroyc's version of *Wuthering Heights* in January 2009.

386. Martín, op. cit., pp. 63-64: 'the intertextual contribution that casting makes to a particular role'.

387. See Sally Wainwright's interview.

388. This popular drama series was broadcast on ITV for four seasons from 2000 to 2003.

389. A phrase taken from the BBC on-line 'press pack'.

390. John Plunkett's 'Sparkhouse Proves to be a Damp Squib' in *The Guardian* (2nd Sept, 2002).

391. Steiner, op. cit., Chap. 5, 'The Hermeneutic Motion', p. 318.

392. The *BBC* made this 'press pack' available on their website on 23rd Aug, 2002. The first episode of the serial was broadcast on BBC1 on Sunday 1st Sept, the second and third episodes both followed on 8th Sept.

    This press pack consists of a single document – a *pdf* file containing text only. This backstage description of *Sparkhouse* is interspersed with some interesting comments from Derek Wax (the producer, Red Productions), from Nicola Schindler (the executive producer, Red Productions), from Gareth Neame (the Head of Independent Commissioning, BBC), from Robin Sheppard (the director), from Sally herself, and from the main members of the cast.

    After the shoot was completed, a series of interviews must have taken place. However, no reference of date(s), place(s) or interviewer(s) is appended to the document.

    http://www.bbc.co.uk/pressoffice/pressreleases/stories/2002/08_august/23/sparkhouse.pdf

393 Two months earlier, Sally responded quite differently on this topic. Here is an excerpt from the interview given by Sally Wainwright and Derek Wax to James Rampton for *The Independent* (Monday 26th Aug, 2002):

> Period adaptations are so boringly literal – and so cynical. *I get annoyed that controllers commission them rather than new work.*
> And, as a writer, I don't know what the challenge would be – all you do is read the book and nick the best lines from the person who wrote it. What's the point? *Straight adaptations are very restricting and very frightening* [...]. [Emphasis added]

Given the terse written answers that Sally gave me, it is reasonable to believe that she chose not to respond fully to my questions about her 'cultural adaptation' of *Wuthering Heights*. Although I had suggested to Sally a phone or a face-to-face interview, the contact I had with her was restricted to a written correspondence via e-mail: there was no opportunity for a live exchange of ideas, and I was unable to reach a stage where I could articulate my appreciation of her work.

In the interview she gave to James Rampton, she seems to have been in the right setting and the right frame of mind to communicate her views on film (and television) adaptations.

394. In the BBC press pack, Robin [Sheppard, the director] explains that:

> It was really important to keep everything real. West Yorkshire was where Sally imagined the story to take place so that's where we shot it. We did not use any sets at all.

395. Sally added in *The Independent* interview (Monday 26th Aug, 2002):

> 'In the novel, Heathcliff is such an evil bastard, he's irredeemable', the writer says. 'Whenever I read it, I come away hating him. But in *Sparkhouse*, I wanted to explain why Carol is so damaged. I hope she emerges as more likeable than him.'

396. This popular drama series was broadcast on ITV for four seasons from 2000 to 2003.

397. See the uncredited article of *The Brighouse Echo* (14th Oct, 2002):

> The Smith Art Gallery in Brighouse hosted a hugely successful opening of its current exhibition, 'Peter Brook in the Pennines'. It was officially opened by Sally Wainwright, writer of BBC1's primary drama for the autumn season, Sparkhouse, accompanied by one of its stars, Sarah Smart, and Robin Sheppard [...]
>
> Sally's connection with Peter Brook goes back many years to his career as a teacher and her days as a student.
>
> In her quest for suitable film locations for Sparkhouse she [Sally] drew on Peter's extensive knowledge of Pennine landscapes and farm buildings. She eventually chose locations around Hebden Bridge, only a few miles from her own roots, and in the next valley to Top Withens, the ruined building which was the inspiration for Emily Brontë's Wuthering Heights.
>
> Peter then used the Sparkhouse theme for a series of paintings created for Sally and featuring several locations used in the TV drama. The principal one is a portrait of Sally herself in front of the Sparkhouse farmhouse and several of the 'Den', the ruined meeting place of the young lovers in Sally's story.

398. See *Wuthering Heights*, Vol. I, Chap. VI, p. 49, the speech spoken by Heathcliff to Nellie:

> I'd not exchange, for a thousand lives, my condition here, for Edgar Linton's at Thrushcross Grange – not if I might have the privilege of flinging Joseph off the highest gable, *and painting the house-front with Hindley's blood*! [Emphasis added]

399. Rivette's interview.
400. Marc Chevrie's article 'La main du fantôme' in *Cahiers du Cinéma* (October 1985).
401. Fabienne Babe as Catherine/Cathy; Lucas Belvaux as Roch/Heathcliff; Sandra Montaigu as Hélène/Ellen, the servant; Olivier Cruveiller as Guillaume/Hindley; Alice de Poncheville as Isabelle/Isabella; and Olivier Torres as Olivier/Edgar.
402. Read *Wuthering Heights*, Vol. II, Chap. III, Dr Kenneth's half-comforting words to a distraught Ellen as he announces Hindley's death to her:

> Hindley Eanshaw! Your old friend Hindley [...] He's barely twenty-seven, it seems; that's your own age; who would have thought you were born in one year!

403. The Rivette DVD collector pack released by ARTE in the Autumn of 2002.
404. The 'avance sur recettes' or 'advance on takings' corresponds to a grant given by the CNC (Centre National de la Cinématographie) to film-makers in France. Back in 1985, half of the French film production benefited from this grant that is attributed after examination of the film project (synopsis and scenario, duration, format, location etc) by a commission de sélection or 'jury'. However, the prestige of the director and personalities of the actors, technicians and screenwriters involved in the project usually play a part into the selection process. See Herve Le Roux, 'Un pas en arrière, deux pas en avance', published in *Cahiers du Cinéma* (May 1985), pp. 371-372.
405. Bill Marshall's *André Téchiné* (2007), Chap. 1, p. 27.
406. Josiane Pinon-Kawatake's Interview with Yoshida Kijû (Yoshishige) in *Positif*, no. 312 (1987), pp. 23-26.
407. Kamilla Elliott's 'Literary Cinema in the Form/Content Debate' (Chap. 5), pp. 154-155 and p. 172.
408. Steiner, op. cit., pp. 318-319.
409. Steiner, op. cit., pp. 318-319.
410. Bataille, op. cit., pp. 9-10.
411. Antoine de Baecque's '*Onimaru*: Les dieux et les hommes' in *Cahiers du Cinéma*, no. 413 (1988) pp. 52-53.
412. De Baecque, loc. cit.
413. Bataille, op. cit., p. 4.

414. 1969.
415. Philippe Jacquier and Marion Pranal's *Gabriel Veyre, opérateur Lumière* (1996).
416. Yoshida studied French literature at the Todai University of Tokyo and did his Doctorate thesis on Jean-Paul Sartre.
417. *Women in the Mirror*, in which the theme of the atomic bomb is central, came a decade later.
418. This DVD collector's box does not seem to be available as yet. But *Women in the Mirror* (2001) was released in 2009 with a much earlier film, *Adieu, Clarté d'été* (1968).
419. It was Kazuo Ishiguro's first novel back in 1982.
420. Mariko Okada is a remarkable Japanese actress (and producer) who started her film career in the 1950s with the likes of Mikio Naruse and Yasujiro Ozu.

     Back in 1961, she asked Yoshida to adapt a novel by Shinji Fujiwara for her. It was *Akitsu Springs* (*Akitsu Onsen*, 1962), her hundredth film, which would turn out to be their biggest success at the box office. It is a love story, melodramatic, bold and totally captivating.

     Yoshida and Okada have been a couple since their collaboration on *Akitsu Springs*.
421. *Confessions among Actresses* (*Kokuhakuteki Joyuron*, 1971).
422. The University of Michigan.
423. The Shochiku Studios.
424. *The End of Summer* (*Kohayagawa ke no aki*, 1961).
425. *Late Autumn* (*Akibiyori*, 1960) and *An Autumn Afternoon* (*Sanma no Aji*, 1962).
426. Tokihiko Okada.
427. *Akitsu Onsen* (1962).
428. *Mizu de Kakareta Monogatari* (1965).
429. *Onna no Mizuumi* (1966).
430. *Joen* (1967).
431. In the first part of his essay entitled 'Eroticism is the approval of life up until death', Georges Bataille writes:

     I believe eroticism to be the approval of life, up until death. Sexuality implies death, not only in the sense in which the new prolongs and replaces that which has disappeared, but also in that the life of the being who reproduces himself is at stake. [...] Individual death is but one aspect of the proliferative excess of being. [...] The basis of sexual effusion is the negation of the isolation of the ego which only experiences ecstasy by exceeding itself, by surpassing itself in the embrace in which the being loses its solitude. Whether it is a matter of pure eroticism (love-passion) or of bodily sensuality, the intensity increases to the point where destruction, the death of the being, becomes apparent. What we call vice is based on this profound implication of death. And the anguish of pure love is all the more symbolic of the ultimate truth of love as the death of those whom it unites approaches them and strikes them.

To no mortal love does this apply as much as to the union between the heroes of Wuthering Heights. [Emphasis added]

432. 1992.

433. *Hiroshima, mon amour* (1959).

434. See Xan Brooks' article 'Andrea Arnold Finds New Depths in *Wuthering Heights*' in *The Guardian* (Tuesday 6th Sept, 2011) where Arnold's answers to the journalists, at the press conference following the screening of her film, feature prominently:

> Every film is like a journey and this one was longer than most [...] It's been a very difficult film in every way. At times it was almost like the film had a curse on it.

435. Steiner, op. cit., Chap. 5 'The Hermeneutic Motion', p. 317.

436. Jason Solomons' article, 'Venice Film Festival: Britain's Big Splash at the Lido' in *The Observer* (Sunday 11th Sept, 2011).

437. Gilbey's article 'Andrea Arnold: I don't do Easy Rides' in *The Guardian* (Monday 31st Oct, 2011).

438. Another Jason Solomons' article 'Steve McQueen: I Could Never Make American Movies – They Like Happy Endings' in *The Observer* (Sunday 8th Jan, 2012).

439. Nadia Attia's article 'Curzon Interview: Andrea Arnold (ONLINE EXCLUSIVE)' in *Curzon Cinemas Online Magazine* (29th Sept, 2011).

440. *Wuthering Heights*, Vol. I, Chap. II, p. 11:

> Her position before was sheltered from the light: now, I had a distinct view of her whole figure and countenance.

441. Steiner, op. cit., Chap. 5 'The Hermeneutic Motion', p. 376.

442. Philip French's '*Wuthering Heights* – Review' in *The Observer* (Sunday 13th Nov, 2011).

443. Xan Brooks' '*Wuthering Heights* – Review' in *The Guardian* (Tuesday 6th Sept, 2011).

444. Philip French's '*Jane Eyre* – Review' in *The Observer* (Sunday 11th Sept, 2011), and '*Wuthering Heights* – Review' in *The Observer* (Sunday 13th Nov, 2011).

445. See pp. 88-93, Vol. I, Chap. IX and X.

446. See pp. 113-119, Vol. I, Chap XI.

447. Andrea Arnold in her interview with Ryan Gilbey entitled, 'Andrea Arnold: I don't do Easy Rides' in *The Guardian* (Monday 31st Oct, 2011).

448. *Wuthering Heights*, Vol. I, Chap. III, p. 20.

449. Gilbey, op. cit.

450. In Barthes' narratological lexicon.
451. The spectral apparitions of Helen Burns, and the more horrifying episodes featuring Bertha ripping Jane's wedding veil or uttering her screams through the little Adèle, only feature in the deleted scenes of the DVD Bonus. They may well appear later in the Director's Cut, once the commercial hurdle of the PG-13 is no longer in the way of a more personal editing.
452. Peter Bradshaw's 'Jane Eyre – Review' in The Guardian (Thursday 8th Sept, 2011).
453. Philip French's 'Jane Eyre – Review' in The Observer (Sunday 11th Sept, 2011).
454. As to Michael Fassbender, his fresh impersonation of the dark and tormented Edward Fairfax Rochester, got him the Best actor Award, from the Los Angeles Film Critics Association Awards (2011).
455. See Peter Bradshaw's 'Jane Eyre – Review' in The Guardian (Thursday 8th Sept, 2011):

> This adaptation is balanced, crafted, beautifully acted, though for me without the thunderclap and lightning-bolt of passion.

456. Solomon Glave, the fourteen-year-old Heathcliff, was also awarded the Young Actors Prize at the Valladolid International Film Festival (2011).
457. Philip French's 'Wuthering Heights – Review' in The Observer (Sunday 13th Nov, 2011).
458. Wuthering Heights, Vol. II, Chap. XVIII, p. 315, and Chap. XIX, pp. 321-322.
459. Bataille, op. cit., pp. 9-10.
460. See the Preface.
461. David Sexton's 'Wuthering Heights – Review' in The London Evening Standard (Friday 11th Nov, 2011).
462. Joseph Paterson's 'Why Wuthering Heights Gives Me Hope' was published in The Guardian on the release date of the film in the United Kingdom, Friday 11th Nov, 2011. Two days later, Steve Rose's 'How Heathcliff Got a Racelift' would be published in the same newspaper.